FOOD LABELING
COMPLIANCE REVIEW

Fourth Edition

FOOD LABELING COMPLIANCE REVIEW

Fourth Edition

James L. Summers

With contributions by
Elizabeth J. (Betty) Campbell

Blackwell Publishing

Author: James L. Summers is Senior Consultant with EAS Consulting Group, LLC (Alexandria, VA), a leading provider of regulatory services to the food, dietary supplement, and cosmetic industries. He is a former FDA food labeling expert with 32 years tenure at FDA. He has held positions as Aquatic Sampling Specialist, Supervisory Microbiologist, Public Health Sanitarian, General Biologist, FDA Inspector, Regional Shellfish Specialist, and Consumer Safety Officer (Division of Regulatory Guidance). His most recent positions with the FDA were as Supervisory Consumer Safety Officer, Branch Chief, and Senior Consumer Safety Officer in the Office of Food Labeling, where he handled the most controversial, complex, and precedent-setting problems involving regulatory compliance issues dealing with food labeling. He participated in the development of policies and regulatory strategies regarding the enforcement of NLEA and other food labeling regulations.

Contributor: Elizabeth J. (Betty) Campbell now serves as Vice President of EAS Consulting Group, LLC (Alexandria, VA), following a 35-year career with the FDA where she worked as Director of Programs and Enforcement Policy in the Office of Food Labeling in the Center for Food Safety and Applied Nutrition and as Acting Director of the Office of Food Labeling. Ms. Campbell played a key role in writing the Nutrition Labeling and Education Act (NLEA) regulations in the early 1990s, and then had major responsibility for implementing those regulations.

©2007 James L. Summers
All rights reserved

Blackwell Publishing Professional
2121 State Avenue, Ames, Iowa 50014, USA

Orders: 1-800-862-6657
Office: 1-515-292-0140
Fax: 1-515-292-3348
Web site: www.blackwellprofessional.com

Blackwell Publishing Ltd
9600 Garsington Road, Oxford OX4 2DQ, UK
Tel.: +44 (0)1865 776868

Blackwell Publishing Asia
550 Swanston Street, Carlton, Victoria 3053, Australia
Tel.: +61 (0)3 8359 1011

Authorization to photocopy items for internal or personal use, or the internal or personal use of specific clients, is granted by Blackwell Publishing, provided that the base fee is paid directly to the Copyright Clearance Center, 222 Rosewood Drive, Danvers, MA 01923. For those organizations that have been granted a photocopy license by CCC, a separate system of payments has been arranged. The fee code for users of the Transactional Reporting Service is ISBN-13: 978-0-8138-2181-8/2007.

First edition, © 1999, AAC Consulting Group, Inc.
Second edition, © 2001, AAC Consulting Group, Inc.
Third edition, © 2003, Iowa State Press

Library of Congress Cataloging-in-Publication Data

Summers, James L. (James Lee), 1936-
   Food labeling compliance review / James L. Summers. — 4th ed.
      p. cm.
   Includes bibliographical references and index.
   ISBN-13: 978-0-8138-2181-8 (alk. paper)
   ISBN-10: 0-8138-2181-9 (alk. paper)
   1. Food—Labeling—United States.   I. Title.

TX551.S83 2007
363.19920973—dc22
                                        2007008020

The last digit is the print number: 9 8 7 6 5 4 3 2 1

Preface

The *"Food Labeling Compliance Review"* publication is a comprehensive compliance food labeling guide designed to aid in understanding the requirements of the Food and Drug Administration (FDA) as it relates to foods other than dietary supplements. This invaluable tool can assist regulatory officials, industry personnel, and others responsible for assuring that the label and labeling of food products in interstate commerce comply with the requirements of the Federal Food, Drug, and Cosmetic Act, as amended, and its regulations.

The publication, authored by James L. Summers, former FDA food labeling expert in the Office of Food Labeling, is a practical, hands-on, user-friendly tool that contains information based on provisions published in the Federal Food, Drug, and Cosmetic Act, as amended; the regulations issued under the act, and FDA's policies and interpretations.

Disclaimer

The information contained in this manual is accurate to the best of the author's knowledge. However, laws, regulations, policies, and official interpretations are subject to change. Since conditions under which the information presented in this publication is used are beyond the control of the author and publisher, we assume no responsibility for the usage of this information. Therefore, no expressed or implied warranty or guarantee should be construed from the content of this publication or its periodic revisions.

Acknowledgment

The publication is dedicated to my devoted wife, Annie B. Summers, and my three daughters, Vilecia C. Summers, Kiea Y. Robertson, and Narvia M. Summers.

Special thanks are also given to Edward A. Steele and Elizabeth J. Campbell of EAS Consulting Group, Inc., for their significant contributions toward bringing this project to a successful conclusion.

Contents

Contents

Contents

Contents

Chapter I
Introduction

The mission of the Food and Drug Administration (FDA) is to enforce laws enacted by the U.S. Congress and regulations promulgated by the Agency to protect the consumer's health, safety, and pocketbook.

The Federal Food, Drug, and Cosmetic Act (FD&CA) is the basic food and drug law of the United States. With numerous amendments, it is the most extensive law of its kind in the world. Many of the individual States have laws similar to the Federal law, and some have provisions to automatically add any new Federal regulations established under it.

The FD&CA is intended to assure the consumer that foods are pure and wholesome, safe to eat, produced under sanitary conditions, that all labeling and packaging is truthful, informative, and not deceptive. Other major acts concerned with the labeling of food products are the Fair Packaging and Labeling Act (FPLA), which affects the content and placement of information required on the packaging, and the Nutrition Labeling and Education Act of 1990 (NLEA), that amended the FD&CA and requires nutrition labeling for most foods. The NLEA also authorizes the use of nutrient content claims and appropriate FDA-approved health claims on the label and labeling of food products.

The purpose of the NLEA is simple: to clear up the confusion that has prevailed on supermarket shelves for years, to help consumers choose more healthful diets, and to offer an incentive to food companies to improve the nutritional qualities of their products. The implementation of the NLEA has been described as embodying *"sweeping reform," "a major overhaul,"* or *"revolutionary change"* in the food label. The enactment of the NLEA fundamentally altered the governing philosophy and legal rules of the game for providing health-related information on food labels. In a speech, Michael Taylor, then Deputy Commissioner for Policy at FDA, stated that "the fundamental philosophical and legal shift embodied in the NLEA...is that a major share of the discretion previously enjoyed by food marketers in deciding the appropriate use of various nutrient descriptors and claims has been transferred from the food companies to FDA. This is the core, unavoidable legal reality of NLEA. Henceforth, the presence of nutrition labeling on packaged foods will not be left to the discretion of the food company. It will be mandatory."

The FD&CA prohibits the distribution within or importation into the United States of articles that are adulterated or misbranded. The term *"adulterated"* refers to products that are defective, unsafe, filthy, or produced under unsanitary conditions, but also includes economic adulteration which would involve product labeling. *"Misbranding"* includes statements, designs, or pictures in labeling that are false or misleading, as well as failure to provide required information on the label or in labeling.

FDA is responsible for protecting the integrity of the food label. The Agency has advised that it intends to evaluate any health claims that appear in labeling on a case-by-case-basis, and that it is prepared to take action against products that make false or misleading health claims.

On January 6, 1993, FDA published 26 final rules addressing various aspects of food labeling. Publication of these final rules culminated in an effort for labeling reform, initiated by the Secretary of Health and Human Services in 1989, and the enactment of the NLEA. The requirements of these regulations have resulted in revision of almost all food labels. This extensive modification of the regulations and the requirement that products shipped in interstate commerce comply with the new requirements have stretched FDA's label review force to monitor and audit products for compliance with these labeling requirements. With this additional workload and the downsizing of most governmental agencies, J. Summers Associates, Inc., has developed this review tool *(Food Labeling Compliance Review)* to facilitate the review process.

This book is not intended to take the place of either the laws or the regulations. It is intended to serve as an aid in understanding the FD&CA as amended, the regulations established under this act as well as those under FPLA. It also serves as a useful tool in assisting regulatory officials and others responsible for the manufacture, labeling, relabeling, and distribution of food products in their review of such labels and labeling in interstate commerce.

This publication consists of a series of established requirements, in the form of questions and responses to be used by the reviewer in establishing the degree of compliance of a food's label and labeling with applicable laws and regulations. It also provides a basis for developing a response concerning the labeling deviations observed during the review and advising the responsible firm about labels needing corrections.

This publication can:

1. Simplify the review process, by systematically asking appropriate questions to affect an effective compliance review;

2. Serve as a teaching aid for less experienced compliance officers and as a ready reference for experienced compliance officers;

3. Improve the accuracy of the review by providing pertinent information for consideration during the review;

4. Provide consistency among reviewers by standardizing the review process (i.e., to help federal label review officers, headquarters and field personnel, and state and local officials, etc., to conduct label reviews in the same manner);

5. Provide consistency among the responses to the responsible firm by using standardized language provided for items needing correction;

6. Shorten management review time because specific items and/or sections can be targeted for review rather than having to review every item;

7. Provide increased and more consistent updating of priorities and changes in regulatory initiatives, and greater coordination with other allied units; and

8. Provide ready access to preamble and regulation citations, as well as helpful illustrations and charts, to assist in the review of labels and in answering inquirers' questions.

This tool also provides the flexibility to enable the user to proceed through the complete review or to review certain components of the label, as needed. Using this book will result in increased productivity and resource savings.

Chapter II
Overview of the History of Food Labeling

Section A:

1. **Any history of food labeling probably should start with the first federal law on food–the Federal Food and Drugs Act of 1906.**

   a) Food content and food claims were not really important issues in 1906.
   b) The law contained no specific requirement that the content of food be disclosed.
   c) It stated only that any declaration on the label of a food had to be truthful and not misleading. Significantly, statements or claims about a food made anywhere but on the label were not subject to the act.
   d) In 1906, the States were predominant in food regulation. The primary question was what role, if any, the federal government should play.

2. **Congress did address the issue of food content in the Federal Food, Drug, and Cosmetic Act (FD&CA), which was passed in 1938, but it did not find a great need to describe food content to consumers.**

   a) Congress took the view that the content of most foods was well known to consumers. Foods like bread, milk, and cheese were made using well-known recipes, which Congress directed the agency to codify as food standards.
   b) If a food bore one of these standardized names, everyone would know what it contained, so an ingredient list was considered unnecessary.
   c) The only lists that Congress specified were for optional ingredients in standardized foods and ingredients for foods not subject to a standard.

3. **As for claims, Congress found that the loopholes in the 1906 Act were widely abused.**

   a) Thus, Congress added labeling–that is, any written, printed, or graphic matter that accompanies the food–to its regulatory scope.
   b) In addition, Congress defined "misleading" in the law. Under this definition, food labeling could be misleading not only for what it said, but also for what it failed to say. The law states that in deciding whether labeling is misleading, the Agency must consider whether it fails to disclose information that is material in light of the representations that are made or of the consequences of consuming the food.

4. **In the wake of the 1938 Act, and well into the 1970's, the information required on a food label was limited.**

   a) Most food complied with a food standard, so there was little ingredient labeling. Those foods that resembled a standardized food, but didn't comply with its standard, risked being labeled as an imitation, with the accompanying pejorative implications.
   b) Other than foods for special dietary use, few foods had labeling disclosing their nutrient content.
   c) Any foods whose labeling mentioned a disease were considered to be drugs.
   d) For example, in 1957, an American Heart Association report stated that dietary saturated fat and cholesterol intake from food could affect the risk of developing heart disease.
   e) FDA considered regulations that would have required disclosure of the level of these food components but ultimately decided that the evidence was not strong enough to establish the existence of a relationship between these components and the disease.

5. **The Fair Packaging and Labeling Act (FPLA) was passed in 1966. Its main purpose was to ensure that consumers had accurate information about the quantity of a product in a package, and could, therefore, make proper price comparisons between products.**

Section B:

1. **Things began to change, however, with the White House Conference on Food, Nutrition, and Health in 1969.**

   a) The conference was held in response to problems of hunger and malnutrition in America.
   b) Its report criticized FDA's restrictive approach to food labeling.
   c) It emphasized that modern food technology had the capability to fill the need for sound nutrition, and that food labeling could provide increased information about nutrition.

2. **FDA responded to the Conference report in several ways.**

   a) The most significant ways were with respect to food content.
      (i) The Agency began to take a much narrower view of what foods were imitations of standardized foods.
      (ii) It was no longer enough that a food resembled a standardized food. Under the Agency's new view, it also had to be nutritionally inferior.
      (iii) More foods were subject to the ingredient labeling requirements of the Act.
      (iv) In addition, FDA began to urge food manufacturers to list both mandatory and optional ingredients on their labels.
      (v) FDA also began to amend its food standards to convert as many mandatory ingredients as possible to optional ingredients, then requiring that those optional ingredients be listed on the label.
   b) A second major development with respect to the nutrient content of a food occurred in 1973. FDA adopted a regulation that provided for the nutrition labeling of food. Under this regulation, if any nutrient is added to a food, or if the label or labeling makes any nutrition claim, the product must bear full nutrition labeling.
      (i) FDA stated that, in these circumstances, the nutrition labeling is a material fact.
      (ii) In other words, if a claim is made about a nutrient in a food, then the full nutrition profile of the food is a material fact that has to be disclosed if the food is not to be misbranded.
      (iii) Consistent with the concerns of the White House Conference, the nutrition label that FDA specified emphasized vitamins and minerals, but also included calories, protein, fat, and carbohydrates.
   c) While FDA greatly expanded the food content information that would be included on food labels, its view of claims remained quite narrow.
      (i) For example, around this time FDA adopted regulations that defined the terms "low calorie" and "reduced calorie."
      (ii) Significantly, the Agency did not do so to allow individuals in the general population to control their calorie intake.

      (iii) Rather, it did so because it felt that such terms would be of special dietary value to those who were controlling their weight on the advice of a doctor.

    d) Similarly, FDA revisited the issue of what information could be provided about the saturated fat and cholesterol contents of food.

      (i) Again, FDA saw no value in providing information about these nutrients to the general population.

      (ii) It provided information about the amount of saturated fat or cholesterol in a food that could be provided in the nutrition label. However, a statement had to accompany the information reporting that it was being provided only for the information of those individuals who were controlling their intake of these nutrients on the advice of a doctor.

    e) As for claims about disease, FDA still held the strict position that such claims made the product a drug.

3. **To close out the 1970's, FDA held joint hearings with the U.S. Department of Agriculture (USDA) and the Federal Trade Commission (FTC) on what changes were necessary in food labeling – what emerged was an agenda for the 1980s.**

    a) With respect to content information, this agenda pointed out that although a growing number of foods had ingredient and nutrition labeling, neither was required. FDA indicated that it intended to seek legislative authority to require such labeling on all foods.

    b) As for claims, FDA announced that it appeared that regulations defining sodium and cholesterol claims would be useful to the general population. It also announced it would study the usefulness of information on other food components, such as fiber. However, FDA reported again that claims about disease did not belong on the food label.

Section C:

1. **In the early 1980s, FDA's resources were significantly reduced.**

2. **As a result of this reduction:**

    a) FDA was unable to follow the rulemaking agenda on nutrition labeling that it set for itself.

      (i) Thus, in the 1980s FDA was able to produce only one final rule on sodium claims and one proposal on cholesterol claims from this agenda.

      (ii) The sodium final rule is significant, because it defined claims, i.e., "low sodium" and "reduced sodium" for foods for the general population and not just for foods for special dietary use.

    b) Most of FDA's efforts were implemented in food safety matters. The Agency did not seek out cases for enforcement of the so-called "economic requirements" such as those for food labeling.

3. **However, Americans' interest in nutrition and the beneficial properties of foods was growing rapidly. Not surprisingly, industry responded. In 1984, some firms began to make label statements about the level of fiber their products contained and began to associate that fiber with an effect on cancer. Although many in the Agency thought these products were misbranded, and perhaps drugs, FDA did not act immediately against the products, which only encouraged more of such claims.**

4. **In 1987, the FDA's proposal on health claims essentially reported that these claims, previously considered by the Agency to be drug claims, could be made on food labels and labeling. In addition, even though it was a proposal, manufacturers could start making claims in accordance with it.**

    a) At this point, FDA's control over food labeling claims was all but gone. The supermarket shelves became lined with products bearing claims on the health effects of their ingredients, some were valid, many were not.

b) The States, whose role in the protection of the food supply had been shrinking gradually since 1938, indicated that if FDA was not going to do anything, they would. A number of State Attorneys Generals got together and began to take aggressive actions against food products.
c) This led to concern in industry, and to reconsideration of FDA deregulation.
d) In 1989, both the Surgeon General and the National Academy of Sciences issued major reports that showed evidence accumulated over the years and revealed that diet could have a significant effect on the risk of developing certain chronic diseases.

5. The results in 1989 were:

a) The food label was out of control;
b) FDA appeared to be powerless;
c) Consumers wanted the situation controlled;
d) The States were becoming active;
e) The Food Industry, concerned about the States, expressed a willingness to compromise; and
f) Science showed that consumers would benefit from healthful dietary practices.

Section D:

1. Pre-NLEA: Congress began to consider whether legislation was necessary. FDA, concerned about how it would be dealt with on Capitol Hill, already had several regulatory initiatives under development.

a) In August of 1989, FDA published an advanced notice of proposed rulemaking (ANPR) on a wide range of food labeling issues, many of which where drawn from the notice that FDA published with USDA and the FTC in 1979. This ANPR touched off a flurry of FDA activity. In November and December of 1989, the Agency held a series of four public hearings around the country on food labeling issues.
b) In February of 1990, FDA published a re-proposal on health claims and withdrew the earlier 1987 proposal that dealt with the distinction between foods and drugs.
c) In March of 1990, the Department of Health and Human Services announced a major departmental food labeling initiative.
d) In July of 1990, FDA published proposals to require nutrition labeling on virtually all foods, to define serving sizes for foods, and to update the U.S. RDAs and establish a series of reference values for macronutrients called Daily Reference Values, as well as a tentative final rule on cholesterol claims.
e) By October of 1990, FDA developed new documents on cholesterol and saturated fat claims. However, this effort did not go far enough.

2. In October of 1990, Congress passed the Nutrition Labeling and Education Act of 1990 (NLEA), which was signed in November of that year by President Bush.

a) The NLEA was enacted to address the following three basic issues involving food labeling that had developed over the preceding 25 years.
   (i) How should the content of food be disclosed on the food label, and how much information about content should be provided?
   (ii) What claims could be made in food labeling?
   (iii) What role should the States have in protecting the public against misbranded food?
b) In significant ways, the NLEA was shaped by the food labeling history described above.
   (i) In the 1989 ANPR the FDA placed the issues of how to describe food content back on the table.
   (ii) Although FDA felt it could require nutrition labeling on most foods based on its pre-NLEA authority, Congress removed any doubt about FDA's authority by requiring such labeling under the NLEA.
   (iii) Congress also required specific ingredient labeling on all foods and of virtually all ingredients.
   (iv) As for claims, the NLEA required that all nutrient content and health claims be authorized by FDA before they can appear on food packages. This requirement followed the model

established by FDA in regulations that defined terms or authorized claims. The NLEA also resolved, definitively, that these claims are for the general population, not simply for those with special dietary needs.

(v) Finally, because of the States' involvement in food labeling enforcement in the late 1980s, enforcement and preemption were issues with which Congress felt compelled to deal. To allay industry's concerns about the States, Congress made most FD&CA misbranding provisions preemptive, thereby providing a uniform national standard for food labeling. This provided Industry with some assurance that if it satisfies FDA, it also will satisfy the States.

Section E:

1. Food and Drug Administration Modernization Act (FDAMA) of 1997 (Pub. L. 105-115)

a) On November 21, 1997, the President signed FDAMA into law, which, among other things, amended Section 403 of the FD&CA to provide for the use in food labeling of nutrient content claims and health claims based on authoritative statements. The FDAMA amendments allow manufacturers and distributors to use claims if such claims are based on current, published, authoritative statements from certain federal scientific bodies, as well as from the National Academy of Sciences. The act requires that a notification of the prospective nutrient content claim or the prospective health claim be submitted to FDA at least 120 days before the food bearing the claim is introduced into interstate commerce. Prior to the passage of these amendments, companies could not use such claims unless FDA published a regulation authorizing the claim.

b) The FDAMA amendments to section 403 also removed the requirement that a referral statement be made on food labels whenever a nutrient content claim is made. However, when a claim is made in labeling about a nutrient in the food that occurs at a level that the Secretary has determined, it increases the risk of a disease or health-related condition to the general population, the amendments require that food labels must bear the disclosure statement "See nutrition information for _____ content" (name of nutrient associated with risk), prominently and in immediate proximity to the claim.

Section F:

1. FDA issued a final rule on July 11, 2003 (68 FR 41434) to require food labels to bear the gram amount of *trans* fat without a percent Daily Value (% DV) on the Nutrition Facts panel. The trans fat final rule amended *21 CFR 101.9* Nutrition Labeling of Food at *§ 101.9(c)(2)*. The effective date for the *trans* fat labeling final rule was January 1, 2006. However, manufacturers were encouraged to voluntarily label *trans* fat before January 1, 2006. Any product that was initially introduced into interstate commerce on or after January 1, 2006 must be labeled with *trans* fat. Sources of *trans* fat include partially hydrogenated oil and some animal-based foods.

Section G:

1. As a result of the Food Allergen Labeling and Consumer Protection Act of 2004 (FALCPA), manufacturers are required to identify in plain English the presence of ingredients that contain protein derived from milk, eggs, fish, crustacean shellfish, tree nuts, peanuts, wheat, or soybeans in the list of ingredients or to state "contains" followed by name of the source of the food allergen after or adjacent to the list of ingredients.

Chapter III
Definitions

1. **Classes of Nutrients for Purposes of Compliance**–The classes are as follows:

   a) Class I–Added nutrients in fortified or fabricated foods.
   b) Class II–Naturally occurring (indigenous) nutrients. If any ingredient that contains a naturally occurring (indigenous) nutrient is added to a food, the total amount of such nutrient in the final food product is subject to Class II requirements, unless the same nutrient also is added. *(21 CFR 101.9(g)(3))*.

2. **Common or Usual Name**–May be a coined term; shall accurately identify or describe, in as simple and direct terms as possible, the basic nature of the food or its characterizing properties or ingredients. *(21 CFR 102.5(a))*

   A common or usual name may be established by common usage or by regulation. *(21 CFR 102.5(d))*

3. **Food**–Articles used for food or drink for man or other animals, chewing gum, and articles used for components of any other such article. *(FD&CA, Sec. 201(f))*

4. **Food Allergy**–A reaction of the body's immune system to something in a food or an ingredient in a food–usually a protein. It can be a serious condition and should be diagnosed by a board–certified allergist. (Understanding Food Allergy, International Food Information Council Foundation, August 1998)

5. **Imitation Food**–A food that is a substitute for and resembles another food but is nutritionally inferior to that food. *(21 CFR 101.3(e))*

6. **Immediate Container**–Does not include package liners. *(FD&CA, Sec. 201(l))*

7. **Information Panel**–The panel immediately contiguous to the right of the PDP as observed by an individual facing the PDP. *(21 CFR 101.2(a))*

   a) If the panel to the right is too small to accommodate the necessary information or is otherwise unusable label space, e.g., folded flaps or can ends, the panel immediately contiguous and to the right of this part of the label may be used. *(21 CFR 101.2(a)(1))*
   b) If the package has one or more alternate PDPs, the information panel is immediately contiguous and to the right of any PDP. *(21 CFR 101.2(a)(2))*

c) If the top of the container is the PDP and the package has no alternate PDP, the information panel is any panel adjacent to the PDP. *(21 CFR 101.2(a)(3))*

8. **Insignificant Amount**—The amount that allows a declaration of zero in nutrition labeling, except for the total carbohydrate, dietary fiber, and protein, it shall be an amount that allows a declaration of "less than 1 gram." *(21 CFR 101.9(f)(1))*

9. **Label**—Display of written, printed or graphic matter upon the immediate container of any article. A requirement made by or under authority of the Act that any word, statement, or other information appearing on the label shall not be considered to be complied with unless such word, statement, or the information also appears on the outside container or wrapper, if there be any, of the retail package of such article, or is easily legible through the outside container or wrapper. *(FD&CA, Sec. 201(k))*

10. **Labeling**—All labels and other written, printed, or graphic matter upon any article or any of its containers or wrappers, or accompanying such article. *(FD&CA, Sec. 201(m))*

11. **Measurable Amount**—A measurable amount of an essential nutrient in a food shall be considered to be 2 percent or more of the Daily Reference Value (DRV) of protein listed under *21 CFR 101.9(c)(7)(iii)* and of potassium listed under *21 CFR 101.9(c)(9)* per reference amount customarily consumed, and 2 percent or more of the Reference Daily Intake (RDI) of any vitamin or mineral listed *under 21 CFR 101.9(c)(8)(iv)* per reference amount customarily consumed, except that selenium, molybdenum, chromium, and chloride need not be considered. *(21 CFR 101.3(e)(4)(ii))*

12. **Nutritional Inferiority**—Any reduction in the content of an essential nutrient that is present in a measurable amount, but does not include a reduction in caloric or fat content, provided that the food is labeled pursuant to the provisions of *21 CFR 101.9*, and provided that the labeling with respect to any reduction in caloric content complies with the provisions applicable to caloric content in *21 CFR Parts 101* and *105. (21 CFR 101.3(e)(4)(i))*

13. **Package**—Any container or wrapper in which any food is enclosed for use in the delivery or display of such commodity to retail purchasers, but does not include:

   a) Shipping containers or wrappings used solely for the transportation of any such commodity in bulk or in quantity to manufacturers, packers, processors, or wholesale or retail distributors;
   b) Shipping containers or outer wrappings used by retailers to ship or deliver any such commodity to retail customers if such containers and wrappings bear no printed matter pertaining to any particular commodity;
   c) Containers used for tray pack displays in retail establishments; and
   d) Transparent wrappers or containers that do not bear written, printed, or graphic matter that obscures the required label information. *(21 CFR 1.20)*

14. **Principal Display Panel (PDP)** —Part of the label that is most likely to be displayed, presented, shown, or examined under customary conditions of display for retail sale. *(21 CFR 101.1)* The area of the PDP is determined as follows:

   a) Rectangular package - The entire side of the package; the product of the height times the width of that side. *(21 CFR 101.1(a))*
   b) Cylindrical package - 40% of the product of the height of the container times the circumference. *(21 CFR 101.1(b))*
   c) Otherwise shaped containers - 40% of the total surface of the container (excludes tops, bottoms, flanges, shoulders, and necks of bottles or jars, etc.). *(21 CFR 101.1(c))*

15. **Standardized Food**—A food for which a regulation has been promulgated fixing and establishing, under its common or usual name, a reasonable definition and standard of identity, a reasonable standard of quality, and/or reasonable standard of fill of container to promote honesty and fair dealing in the interest of consumers. *(FD&CA, Sec. 401)*

16. **Substitute Food**–A food that is a substitute for and resembles another food but is not nutritionally inferior to the food for which it substitutes and resembles. Its label bears a common or usual name that complies with the provisions of *21 CFR 102.5* and that it is not false or misleading, or in the absence of an existing common or usual name, an appropriately descriptive term or phrase that is not false or misleading. *(21 CFR 101.3(e)(2)(i) & (ii))*

17. *Trans* **Fat**–Unlike other fats, the majority of *trans* fat is formed when liquid oils are made into solid fats like shortening and hard margarine. However, a small amount of *trans* fat is found naturally, primarily in some animal-based foods. Essentially, *trans* fat is made when hydrogen is added to vegetable oil -- a process called hydrogenation. Hydrogenation increases the shelf life and flavor stability of foods containing these fats. *Trans* fat, like saturated fat and dietary cholesterol, raises the LDL (or "bad") cholesterol that increases your risk for CHD. (FDA, CFSAN, ONPLDS, Trans Fat Now Listed With Saturated Fat and Cholesterol on the Nutrition Facts Label, January 16, 2004; Updated March 3, 2004; Updated January 1, 2006)

Chapter IV
Changes in Food Labeling Regulations Resulting from NLEA

Following the enactment of the Nutrition Labeling and Education Act (NLEA) of 1990, the Food and Drug Administration (FDA) changed a number of its labeling regulations. Subsequently, FDA's Office of Food Labeling (OFL) issued information and interpretations to alert interested parties to the changes, noted below, resulting from these regulations.

Section A: Ingredient Labeling

All ingredients in FDA's standardized foods provided for under Title 21 of the Code of Federal Regulations (CFR) must be declared in the ingredient statement on the food label. Where a standard of identity contains a specific provision with respect to the declaration of optional ingredients, and such declaration is necessary to differentiate between two or more foods that comply with the standard, the ingredients must be declared in accordance with the provisions of that standard. *(21 CFR 130.11)*

Section B: Certified Color Additives

FDA's certified color additives and their lakes must be individually declared by their common or usual names or by the permitted abbreviated names (e.g., *"FD&C Blue No. 1"* or *"Blue 1"* or *"FD&C Yellow No. 5,"* or *"Yellow 5,"* or *"Blue 1 Lake"*). *(21 CFR 101.22(k))*

Section C: Common or Usual Names for Nonstandardized Foods

1. **Protein Hydrolysates:** *(21 CFR 102.22)*

   The common or usual name of a protein hydrolysate must include the food source from which it was derived, e.g., *"hydrolyzed wheat gluten"* or *"hydrolyzed casein."* The names, *"hydrolyzed vegetable protein" and hydrolyzed plant protein"* are not acceptable.

   If the source of proteins that are hydrolyzed is undifferentiated, i.e., the source is not fractionated into distinct proteins but is composed of all proteins naturally occurring in that source, the common or usual name would reflect the whole protein component of the source, e.g., *"hydrolyzed wheat protein" and "hydrolyzed milk protein."*

2. **Beverages that Contain Fruit or Vegetable Juice:** *(21 CFR 101.30 and 102.23)*

   The juice and juice beverage regulations are divided into two parts. One, dealing with percent juice labeling, is in *21 CFR 101.30*, while the other, on establishing common or usual names for beverages that contain juice, is in *21 CFR 102.23*.

Section D: Standardized Foods

FDA regulations pertaining to standardized foods *(21 CFR Parts 130 through 169)* were affected by the FDA's regulatory amendments.

1. **General Provisions:** *(21 CFR Part 130)*

    Part 130 -- Food Standards: General -- Includes sections 130.3, Definitions and interpretations; 130.9, Sulfites in standardized foods; 130.10, Requirements for foods named by the use of a nutrient content claim and a standardized term; and 130.11, Label designations of ingredients for standardized foods. One significant change in this part was the establishment of the general standard in 21 CFR 130.10, which provides for variations in traditional standardized foods to achieve a nutritional benefit, such as a reduced or lower fat content, provided certain conditions regarding nutritional quality, ingredients, and functional characteristics are met.

2. **Specific Varieties of Foods:** *(21 CFR Parts 131 through 169)*

    Changes in these standards of identity, for the most part, provide that each ingredient (mandatory and optional) used in the food must be declared on the label as required by the applicable sections of *21 CFR Parts 101 and 130.*

Section E: Nutrition Labeling

1. **Mandatory Labeling:**

    Nutrition labeling reform was intended to clear up confusion that had prevailed on supermarket shelves for years, to help consumers choose more healthful diets, and to offer an incentive to food companies to improve the nutritional qualities of their products.

    Among the key changes taking place were:

    (a) Mandatory nutrition labeling for almost all foods. Consumers now are able to learn about the nutritional qualities of almost all of the food products they buy.
    (b) Provision for information on the amount per serving of saturated fat, cholesterol, dietary fiber, and other nutrients of major health concern to today's consumers.
    (c) Establishment of nutrient reference values, expressed as Percent of Daily Value, to be used in the labeling so consumers can see how a food fits into the overall diet.
    (d) Establishment of uniform definitions for terms describing a food's nutrient content, such as *"light," "low fat," and "high fiber,"* to ensure that such terms mean the same for any product on which they appear.
    (e) Authorization of health claims about the relationship between a nutrient and a disease, such as calcium and osteoporosis, and fat and cancer.
    (f) Standardization of serving sizes to simplify nutrition comparisons of similar products.
    (g) Declaration of total juice percentage on labels of foods that purport to contain juice, enabling consumers to know exactly how much juice is in a product.
    (h) Provision for voluntary nutrition information for many raw foods.

    These and other changes to the FDA's regulations require nutrition labeling for most foods, and authorize the use of nutrient content claims and appropriate FDA approved health claims. Meat and poultry products regulated by USDA are not covered by NLEA. However, USDA has established regulations that closely parallel those established by FDA under NLEA.

2. **Nutrition Labeling Exemptions:**

    While the new regulations require nutrition labeling on most food products, nutrition labeling is voluntary for many raw foods. For example, under FDA's voluntary point-of-purchase nutrition information program, the 20 most frequently eaten raw fruits, vegetables, and fish may bear nutrition labeling. In fact, point-of-purchase information for raw produce and raw fish has been available in some grocery stores since November of 1991.

Although voluntary, the programs for raw produce and raw fish carry strong incentives for retailers to participate. The NLEA states that if voluntary compliance is insufficient, nutrition information for such raw foods will become mandatory.

Under NLEA, some foods are exempt from nutrition labeling. These include:

(a) Food produced by small businesses (those with food sales of less than $50,000 a year or total sales of less than $500,000), provided that the food bears no nutrition claims or other nutrition information on its label or labeling.
(b) Food products that are low-volume (that is, they meet the requirements for "units sold" in *21 CFR 101.9(j)(18)(i) or (j)(18)(ii)*, in accordance with the requirements of *101.9(j)(18)(iv)*.
(c) Restaurant food.
(d) Food served for immediate consumption, such as that served in hospital cafeterias and on airplanes.
(e) Ready-to-eat food prepared primarily on site; for example, bakery, deli, and candy store items.
(f) Food sold by food service vendors, such as mall cookie counters, sidewalk vendors, and vending machines.
(g) Goods shipped in bulk, as long as it is for sale in that form to consumers.
(h) Medical foods, such as those used to address the nutritional needs of patients with certain diseases.
(i) Plain coffee and tea, some spices, and other foods that contain no significant amounts of any nutrients.
(j) Packages with less than 12 square inches available for labeling do not have to carry nutrition information. However, they must provide an address or telephone number for consumers to obtain the required nutrition information.
(k) Game meats, such as deer, bison, rabbit, quail, ostrich, etc. Nutrition information for these foods may be provided on counter cards, signs, or other point-of-purchase materials, rather than on individual labels. FDA believes that allowing this option will enable game meat producers to give first priority to collecting appropriate nutrition data, and will make it easier for them to update the nutrition information as it becomes available.

3. Nutrient Content Declarations:

The new food label features a revamped nutrition panel. This panel bears the heading *"Nutrition Facts,"* which replaces *"Nutrition Information Per Serving."* The new title will signal to consumers that the product label complies with the new regulations.

The nutrition panel bears new sets of dietary components. The mandatory (underlined) and voluntary components and the order in which they must appear on the label are:

(a) <u>Total calories</u>
(b) <u>Calories from fat</u>
(c) Calories from saturated fat
(d) <u>Total fat</u>
(e) <u>Saturated fat</u>
(f) *Trans* Fat (Added 7/2/03)
(g) Polyunsaturated fat
(h) Monounsaturated fat
(i) <u>Cholesterol</u>
(j) <u>Sodium</u>
(k) Potassium
(l) <u>Total carbohydrate</u>

(m) <u>Dietary fiber</u>

(n) Soluble fiber

(o) Insoluble fiber

(p) <u>Sugars</u>

(q) Sugar alcohol (for example, xylitol, mannitol, sorbitol, and erythritol). Declaration is voluntary unless a claim is made about sugar alcohol or sugars when sugar alcohols are present in food

(r) Other carbohydrates (the difference between total carbohydrates and the sum of dietary fiber, sugars, and sugar alcohol, if declared)

(s) <u>Protein</u> (may be optionally listed in accordance with *21 CFR 101.9(g)(7)*)

(t) <u>Vitamin A</u>

(u) <u>Vitamin C</u>

(v) <u>Calcium</u>

(w) <u>Iron</u>

(x) Other essential vitamins and minerals

If a claim is made about any of the optional components, or if a food is fortified or enriched by them, nutrition information for these components becomes mandatory.

These mandatory and voluntary components are the only ones allowed on the nutrition panel. The listing of single amino acids, maltodextrin, calories from polyunsaturated fat, and calories from carbohydrates, for example, may not appear as part of the Nutrition Facts on the label.

The required nutrients were selected because they address today's health concerns. The order in which they must appear reflects the priority of current dietary recommendations.

Thiamin, riboflavin, and niacin are no longer required in nutrition labeling because deficiencies of these nutrients are no longer considered to be of public health significance. However, they may be listed voluntarily.

4. Nutrition Label Format:

The format for declaring nutrient content per serving also has been revised. Now, most nutrients must be declared as a percent of their Daily Value – the new label reference values. The amount, in grams, of macronutrients (such as fat, cholesterol, sodium, carbohydrates, and protein) still must be listed to the immediate right of the names of each of these nutrients. For the first time, a column headed *"% Daily Value"* will appear, as will a footnote to help consumers meet their individual nutrient needs with respect to the Daily Values used on the label.

The new format requiring nutrients to be declared as a percent of the Daily Value is intended to prevent the misinterpretations that arose with quantitative values.

Some variations in the format of the nutrition panel are allowed, but some are mandatory. For example, the labels of foods for children under 2 years of age (except infant formula, which has special labeling rules under the Infant Formula Act of 1980) may not carry information about saturated fat, polyunsaturated fat, monounsaturated fat, cholesterol, calories from fat, or calories from saturated fat. This is to prevent parents from wrongly assuming that infants and toddlers should restrict their fat intake, when, in fact, they should not. Fat is important during the early years to ensure adequate growth and development.

5. Serving Sizes:

Whatever format is used, the serving size remains the basis for reporting each food's nutrient content. However, as opposed to the past, when the serving size was up to the discretion of the food manufacturer, it now is more uniform and will be expressed in both common household and metric measures.

Chapter IV

NLEA defines serving size as the amount of food customarily eaten at one time. The serving sizes appearing on food labels are based on FDA-established lists of *"Reference Amounts Customarily Consumed Per Eating Occasion."* These reference amounts, which are part of the new regulations, are broken down into FDA-regulated food product categories, including groups of foods especially formulated or processed for infants or children under four years of age. The reference amount for each food is based primarily on national food consumption surveys.

6. **Daily Values - DRVs and RDIs:**

The new label reference value, Daily Value (DV), comprises two new sets of dietary standards: Daily Reference Values (DRVs) and Reference Daily Intakes (RDIs). Only the Daily Value term appears on the label to make label reading less confusing.

The DRVs are used for macronutrients that are sources of energy, e.g., fat, carbohydrates, and protein; and for fiber, cholesterol, sodium, and potassium, which do not contribute to calories.

DRVs for energy-producing nutrients are based on the number of calories consumed per day. A daily intake of 2,000 calories has been established as the reference. This level was chosen because it has the greatest public health benefit for the nation.

The RDI replaces the term *"U.S. RDA" (U.S. Recommended Daily Allowances),* which was introduced in 1973 as a label reference value for vitamins, minerals, and protein in voluntary nutrition labeling. The name change was sought because of confusion that existed over *"U.S. RDAs,"* the values determined by FDA and used on food labels, and *"RDAs"* (Recommended Dietary Allowances), the values determined by the National Academy of Sciences for various population groups and used by FDA to figure the U.S. RDAs. However, the values for the new RDIs remain the same as the old U.S. RDAs for now.

The following RDIs and nomenclature were established for the following vitamins and minerals which are essential in human nutrition. *(21 CFR 101.9(c)(8)(iv))*

(a) Vitamin A, 5,000 International Units
(b) Vitamin C, 60 milligrams
(c) Calcium, 1,000 milligrams
(d) Iron, 18 milligrams
(e) Vitamin D, 400 International Units
(f) Vitamin E, 30 International Units
(g) Vitamin K, 80 micrograms
(h) Thiamin, 1.5 milligrams
(i) Riboflavin, 1.7 milligrams
(j) Niacin, 20 milligrams
(k) Vitamin B_6, 2.0 milligrams
(l) Folate, 400 micrograms
(m) Vitamin B_{12}, 6 micrograms
(n) Biotin, 300 micrograms
(o) Pantothenic acid, 10 milligrams
(p) Phosphorus, 1,000 milligrams
(q) Iodine, 150 micrograms
(r) Magnesium, 400 milligrams
(s) Zinc, 15 milligrams
(t) Selenium, 70 micrograms
(u) Copper, 2.0 milligrams
(v) Manganese, 2.0 milligrams
(w) Chromium, 120 micrograms
(x) Molybdenum, 75 micrograms
(y) Chloride, 3,400 milligrams

7. Nutrient Content Descriptors:

The regulations spell out what terms may be used to describe the level of a nutrient in food and how they can be used. The core terms are: *"free," "low," "lean"* and *"extra lean," "high," "good source," "reduced," "light,"* and *"more."* The regulations also address other claims, such as percent fat free and implied claims.

8. Health Claims:

Claims for relationships between a nutrient or a food and the risk of a disease or health-related condition are allowed for the first time. They can be made in several ways: through third-party references, such as the National Cancer Institute; statements; symbols, such as a heart; and vignettes or descriptors. Whatever the case, the claim must meet the requirements for authorized health claims. For example, they cannot state the degree of risk reduction and can only use *"may" or "might"* in discussing the nutrient or food-disease relationship. They must state that other factors play a role in that disease.

They also must be phrased in such a way that the consumer can understand the relationship between the nutrient and the disease, and the nutrient's importance in relationship to the daily diet. The allowed nutrient-disease relationship claims are: calcium and osteoporosis; dietary lipids and cancer; sodium and hypertension; dietary saturated fat and cholesterol and risk of coronary heart disease; fiber-containing grain products, fruits, and vegetables and cancer; fruits, vegetables, and grain products that contain fiber, particularly soluble fiber, and risk of coronary heart disease; fruits and vegetables and cancer; folate and neural tube defects; sugar alcohol and dental caries; soluble fiber from certain foods and risk of coronary heart disease; soy protein and coronary heart disease; and plant strerol/stranol esters and coronary heart disease.

Section F: Descriptive Claims: Fresh, Freshly Frozen, Fresh Frozen, Frozen Fresh *(21 CFR 101.95)*

The agency defined *"fresh"* because of concern over the term's possible misuse on some food labels. The regulation defines the term *"fresh"* when it is used to suggest that a food is raw or unprocessed. In this context, *"fresh"* can be used only on a food that is raw, has never been frozen or heated, and contains no preservatives (irradiation at low levels is allowed where appropriate). *"Fresh frozen," "frozen fresh,"* and *"freshly frozen"* can be used for foods that are quickly frozen while still fresh. Blanching (brief scalding before freezing to prevent nutrient breakdown) is allowed. Other uses of the term *"fresh,"* such as in *"fresh milk"* (for pasteurized milk) or *"freshly baked,"* are not affected.

Section G: Others: Noncertified Color Additives, Spices and Flavorings

1. FDA recommends that noncertified color additives be declared using terms such as *"artificial color"* and *"color added"* or by the specific name of the color additive used such as *"annatto color."*

2. Spices and flavorings may continue to be declared as such without naming each individual spice or flavoring. Incidental additives and processing aids do not require declaration.

Chapter V
Outline for Compliance Review

Section A: Establish Jurisdiction

1. **Food Subject to the Requirements of the Federal Food, Drug, and Cosmetic Act, as Amended (FD&CA):**

   (a) Food is covered by the regulations administered by the Food and Drug Administration (FDA).
   (b) Product has been introduced or delivered into interstate commerce.

2. **On Requests for Information on a Label Review and Certificate of Free Sale:**

   (a) Determine whether or not the requester is authorized to receive comments on the label or a certificate of free sale.
   (b) Determine if the label is for institutional products (not sold as such at retail) and bears information required on the label by FD&CA.
   (c) Determine if the product is for retail (consumer size container) and bears information required by the FD&CA, the Fair Packaging and Labeling Act (FPLA), the Nutrition Labeling and Education Act (NLEA), and the regulations established under these acts.

Section B: Determine the Completeness and Accuracy of the Required Label Information *(21 CFR Part 101)*

1. **Identity Statement:** *(21 CFR 101.3)*

   The name of the food shall be:

   (a) A principal feature on the principal display panel (PDP);
   (b) The common or usual name of the food; or in the absence thereof;
       (i) An appropriately descriptive term or phrase may be used; or
       (ii) When the nature of the food is obvious, a fanciful name commonly used by the public for such food may be used;
   (c) Accompanied by the particular form of the food, when the food is marketed in optional forms (e.g., whole, slices, dried, etc.), as necessary, to clearly define the product's identity; and
   (d) Presented in bold type, in a size reasonably related to the most prominent printed matter on the PDP, and in a line(s) generally parallel to the base on which the package rests as it is designed to be displayed.

2. **Net Quantity of Contents Statement:** *(21 CFR 101.105)*

    The net quantity of contents shall exclude the package and packing medium that is not consumed.

    (a) The statement is presented on the principal display panel, in terms of weight, measure, numerical count, or combination of numerical count and weight or measure, as follows:
        (i) A statement of weight shall be in terms of avoirdupois pounds and ounces;
        (ii) A statement of fluid measure shall be in terms of U.S. gallon and quart, pint, and fluid ounce subdivisions thereof; and
        (iii) A statement of dry measure shall be in terms of the U.S. bushel and peck, dry quart, and dry pint subdivisions thereof.
    (b) The statement shall appear as a distinct item on the principal display panel separated (by at least a space equal to the height of the lettering used in the declaration) from other printed label information appearing above or below the declaration and (by at least a space equal to twice the width of the letter "N" used in the quantity of contents statement) appearing to the left or right of the declaration.
    (c) The statement shall be placed within the bottom 30 percent of principal display panel, in a line(s) generally parallel to the base on which the package rests.
    (d) The statement shall be in letters and numerals in a type size established in relationship to the area of the PDP.
    (e) The statement shall also be accompanied by a declaration in the SI (metric) units.

3. **Ingredient Statement:** *(21 CFR 101.4)*

    (a) The label of a food with two or more ingredients must list each ingredient as provided for under *21 CFR 101.4* unless the ingredient is exempt from label declaration under *21 CFR 101.100* (incidental additives and processing aids).
    (b) The ingredients are listed in descending order of predominance by their common or usual names, except that ingredients presented in amounts of 2% or less may be declared by the use of a qualifying statement, e.g., contains _____ percent or less of _____ and _____ (the first blank to be filled with the amount and the subsequent blanks to be filled with the names of the ingredients).
    (c) The name of the ingredient shall be a specific name and not a collective or generic name, except that spices, colorings, and chemical preservatives shall be declared as provided in *21 CFR 101.22.*
    (d) An ingredient which itself contains two or more ingredients shall be declared in the statement of ingredients by:
        (i) Declaring the common or usual name of the multi-component food followed by a parenthetical listing of each ingredient contained therein in descending order of predominance by its common or usual name; or
        (ii) Declaring each ingredient in the multi-component food in descending order of predominance in the finished food without listing the ingredient itself.
    (e) The declaration shall be on the principal display panel or on the information panel. When declared on the information panel, it shall appear together with the name and address of the manufacturer, packer, or distributor, and (where applicable) the nutrition information without intervening material.

4. **Name and Address of the Place of Business:** *(21 CFR 101.5)*

    (a) The declaration of the name and address of the manufacturer, packer, or distributor of the product must appear conspicuously on the food label.
    (b) The declaration shall be on the principal display panel or, alternatively, on the information panel together with the ingredient statement and the nutrition information (where applicable).

(c) If the product is not manufactured at the place declared on the label, the declaration must be qualified by a statement indicating the connection the listed firm has with the food, e.g., "*Manufactured for _____*", "*Distributed by _____*", etc.

(d) The street address may be omitted from the declaration, if it is shown in a current local directory or telephone directory.

5. **Nutrition Labeling:** *(21 CFR 101.9)*

Under the NLEA, most food products are required to bear nutrition labeling.

(a) Determine if the product is exempt from nutrition labeling in accordance with the criteria established under *21 CFR 101.9(j)*.

(b) If the product is not exempt from nutrition labeling, examine the label to determine if it bears nutrition labeling in a manner complying with applicable regulations under 21 CFR Part 101, including:

    (i) Location of the Nutrition Facts panel *(21 CFR 101.2(b) and (d))*;

    (ii) Required format *(21 CFR 101.9(d)-(f))*;

    (iii) Appropriate serving size *(21 CFR 101.9(b) and 101.12)*;

    (iv) Declared nutrient values *(21 CFR 101.9(c))*; and

    (v) Nutrient content and health claims *(21 CFR 101.13, 101.14, 101.54 through 101.67, and 101.71 through 101.83)*.

Section C: Determine the Degree of Compliance with Applicable Regulations

Conduct a compliance label review. (See Chapter VI)

Section D: Advise Responsible Firm of Label(s) Needing Corrections

When labeling violations are discovered, there are several regulatory procedures available to effect correction;

1. **Untitled Letter**–An untitled letter is issued in circumstances where the agency needs to communicate with the regulated industry about documented violations that do not meet the threshold of regulatory significance. (Chapter 4 of the *FDA Regulatory Procedures Manual*, August 1995)

2. **Product Recall**–Recall of a violative product from the market by the manufacturer is generally the fastest and most effective way to protect the public. A recall may be voluntarily initiated by the manufacturer or shipper of the product, or at the request of FDA. Guidelines on FDA recall procedures and industry responsibilities are provided under *21 CFR Part 7*.

3. **Warning Letter**–This written communication from FDA serves to notify an individual or firm that the agency considers one or more of its products, practices, processes, or other activities to be in violation of the FD&CA, or other acts, and that failure of the responsible party to take appropriate and prompt action to correct and prevent any future repeat of the violation may result in administrative and/or regulatory enforcement action without further notice. (Chapter 4 of the *FDA Regulatory Procedures Manual*, August 1995)

4. **Seizure**–This is a civil court action against the goods to remove them from the channels of commerce. After seizure, the goods may not be altered, used or moved, except by permission of the court. The owner or claimant of the seized merchandise is usually given about 30 days by the court to decide on his course of action. He may do nothing, in which case the court will dispose of the goods. He may decide to contest the Government's charges by filing a claim and answering the charges and the case will be scheduled for trial; or he may consent to condemnation of the goods, while requesting permission of the court to bring the goods into compliance with the law. In the latter

option, the owner of the goods is required to provide a bond (money deposit) to assure that the orders of the court will be carried out, and must pay for FDA supervision of any compliance procedure. (See Section 304 of the FD&CA)

5. **Citation**–Section 305 Notices are issued pursuant to that section of the FD&CA. Citation may be utilized ONLY when a prosecution recommendation is definitely being considered by the agency. Citation should not be used for warning purposes. (See Chapter 5 of the *FDA Regulatory Procedures Manual*, August 1995)

6. **Prosecution**–Except for prosecution recommendations involving gross, flagrant, or intentional violations, fraud, or danger to health, each recommendation should ordinarily contain proposed criminal charges that show a continuous or repeated course of violative conduct. This may consist of counts from two or more inspections, or counts from separate violative shipments at different points in time. The agency ordinarily exercises its prosecutorial discretion to seek criminal sanctions against a person only when a prior warning or other type of notice can be shown. Establishing a background of warning or other type of notice will demonstrate to the U.S. Attorney, the judge, and the jury that there has been a continuous course of violative conduct and a failure to effect correction in the past. (See Chapter 6 of the *FDA Regulatory Procedures Manual*, August 1995)

7. **Injunction**–An injunction is a civil process initiated to stop or prevent violation of the law, e.g., to halt the flow of violative products in interstate commerce, and to correct the conditions that caused the violation to occur. (See 21 U.S.C. 332; Rule 65, Rules of Civil Procedure, and Chapter 6 of the *FDA Regulatory Procedures Manual*, August 1995)

8. **Release With Comment Notices (Import)**–Release With Comment Notices are considered to be consistent with a "good faith effort" and CFSAN will allow additional time for firms to voluntarily correct their labels. (See Attachment A of the "Guidance For Issuance of Release Notices" in Nutrition Labeling of Imported Foods Assignment, January 10, 1995, and Chapter 9 of the *FDA Regulatory Procedures Manual*, August 1995)

9. **Detention (Import)**–The Notice of Detention and Hearing is a document which charges that an entry of imported merchandise is in violation of the appropriate Act (FD&CA, Public Health Service Act (PHSA), etc.) because it appears to be adulterated or misbranded. It also provides the importer or his designated agent with an opportunity to introduce testimony or to file a statement in writing relative to the admissibility of the lot detained.

It should be noted that the FD&CA does not provide specifically for the issuance of a notice charging that an entry of imported merchandise appears to be in violation. However, *21 CFR 1.94* provides that if it appears that an imported article may be subject to refusal of admission, the FDA district director shall give the owner or consignee a written notice to that effect. (See Chapter 9 of the *FDA Regulatory Procedures Manual*, August 1995)

Chapter VI
Compliance Label Review Program

Section A: Identity Statement (*21 CFR 101.3*)

1. **Summary of the Requirements**

    a) The identity statement shall be one of the principal features on the principal display panel of the label.

    b) The statement of identity shall be in terms of:
        (i)   The name required by any applicable Federal law or regulation; or in the absence thereof,
        (ii) The common or usual name of the food; or, in the absence thereof,
        (iii) An appropriate descriptive term, or when the nature of the food is obvious, a fanciful name commonly used by the public for such food.

    c) When a food is marketed in various optional forms (whole, slices, diced, etc.), the particular form shall be considered to be a necessary part of the statement of identity and shall be declared in letters of a type size bearing a reasonable relation to the size of the letters forming the other components of the statement of identity; except that if the optional form is visible through the container or is depicted by an appropriate vignette, the particular form need not be included in the statement. <u>This specification does not affect the required declarations of identity under definitions and standards for foods promulgated pursuant to section 401 of the act.</u>

    d) When a food is identified by the use of a nutrient content claim and a standardized name, such a food is subject to the requirements of *21 CFR 130.10* (see VI.A.4).

    e) The statement shall be in:
        (i)   Bold type on the principal display panel;
        (ii) A size reasonably related to the most prominent printed matter on the panel; and
        (iii) In a line generally parallel to the base on which the package rests as it is designed to be displayed.

    f) If it is an imitation of another food its label shall bear, in type of uniform size and prominence, the word "*imitation*" and, immediately thereafter, the name of the food imitated. (See VI.A.9)

    g) When the label or labeling makes any direct or indirect representations with respect to a primary recognizable flavor(s), such flavor shall be considered to be the "*characterizing flavor*" and the common or usual name of such flavor shall accompany the identity statement on the principal display panel of the label. (See VI.A.10)

2. Questions and Responses

The questions and responses sections included under Chapter VI are a series of established requirements designed to aid the reviewers in establishing the degree of compliance that a specific food label and its labeling comply with applicable laws and regulations. It is not intended to represent all questions that may arise during the review process, but as examples of typical questions during a label review. This section can also serve as a teaching aid for less experienced reviewers and as a ready reference for experienced reviewers. It can also serve as an example for developing a response to advise those responsible for food labels and labeling of needed corrections.

Is the product in package form? *(21 CFR 101.3(a))*

A. If "*YES*," continue.
B. If "*NO*," the product is not subject to the principal display panel requirements.

Does the label bear a statement of identity on its principal display panel in accordance with 21 CFR 101.3? *(See Illustration VI.A.3 - Principal Display Panel)*

A. If "*YES*," continue.
B. If "*NO*," STOP HERE and state that the product fails to bear an identity statement, within the meaning of section 403(i)(1) of the Act, on the principal display panel as required by *21 CFR 101*.3(a).

Is the statement of identity one that has been established or otherwise provided for by regulation? *(See Regulations and Charts below)*

- Regulations VI.A.4 - Foods Named by Use of a Nutrient Content Claim and a Standardized Term *(21 CFR 130.10)*
- Chart VI.A.6 – List of Standards of Identity for Foods (21 CFR Parts 131 through 169)
- Chart VI.A.7 - Common or Usual Names For Nonstandardized Foods *(21 CFR Part 102)*
- Regulations VI.A.8 - Identity Labeling of Food in Packaged Form *(21 CFR 101.3)*
- Chart VI.A.9 - Requirements for Foods that Resemble and Substitute for Traditional Foods *(21 CFR 101.3(e))*

A. If "*YES*," continue.
B. If "*NO*," state that the food fails to bear an appropriate statement of identity in accordance with _____ (insert citation for the applicable regulation).

Is the food marketed in various optional forms (e.g., whole, slices, diced, etc.) and labeled in accordance with 21 CFR 101.3(c)?

[*NOTE: Under 21 CFR 101.3(c), when the optional form is considered a necessary part of the statement of identity, the form must be declared in a type size reasonably related to the size of the letters in the statement of identity.*
- *The optional form is not required to be declared if the product is visible through the container or the product is depicted by an appropriate vignette on the label.*
- *The optional form exemption does not affect the required declarations of identity under the definitions and standards for foods under section 401 of the FD&CA.]*

A. If "*YES,*" continue.

B. If "*NO,*" but the product should be marketed in optional forms, state that the statement of identity fails to bear the optional form of the product on the principal display panel in accordance with *21 CFR 101.3(c).*

If "*NO,*" because it complies with the *21 CFR 101.3(c)* exemption, or because it is not required to be marketed in optional form, continue.

 Is the statement of identity displayed on the principal display panel in lines generally parallel to the base on which the package rests as it is designed to be displayed? *(21 CFR 101.3(d))*

A. If "*YES,*" continue.

B. If "*NO,*" state that the statement of identity fails to appear on the principal display panel in lines generally parallel to the base on which the package rests.

 Is the statement of identity one of the principal features on the principal display panel? *(21 CFR 101.3(a))*

A. If "*YES,*" continue.

B. If "*NO,*" state that statement of identity is not one of the principal features on the principal display panel.

 Is the statement of identity presented in bold type on the principal display panel? *(21 CFR 101.3(d))*

A. If "*YES,*" continue.

B. If "*NO,*" state that statement of identity is not presented on the principal display panel in bold type.

 Is the statement of identity declared on the principal display panel in a size reasonably related to the most prominent printed matter on the panel? *(21 CFR 101.3(d))*

A. If "*YES,*" continue.

B. If "*NO,*" state that the statement of identity is not declared on the principal display panel in a size reasonably related to the most prominent printed matter on the panel.

 Does the label, labeling, or advertising of the food make any direct or indirect representations with respect to the primary recognizable flavor(s)?

A. If "*YES,*" the flavor labeling should be in accordance with the instructions prescribed for the particular food types. (See Chart VI.A.10) For example:
   - Labeling of standardized foods should be guided by the instructions provided in the standard of identity for the particular food. *(21 CFR Parts 131 through 169);*
   - Labeling of nonstandardized foods, with established common or usual names, should be guided by the regulations for the specific food *(21 CFR Part 102);* and
   - Labeling of other nonstandardized foods should be guided by the regulations in *21 CFR 101.22.*

B. If "*NO,*" STOP HERE. A declaration of flavor as part of the statement of identity is not required.

 Is the recognizable flavor(s) declared on the label in accordance with the applicable requirements above?

A. If "*YES,*" continue.

B. If "*NO*," state that the food makes representations with respect to a recognizable flavor(s), but the label fails to declare such flavor(s) as required by _____ (insert citation for the applicable regulation).

Q

Does the flavor declaration make appropriate reference as to whether the flavor is "natural" or "artificial"?

A. If "*YES*," continue.
B. If "*NO*," state that the label for the product represents its characterizing flavor to be, e.g., a natural flavor as defined by *21 CFR 101.22(a)(3)*, but the _____ (insert evidence of the findings, e.g., formula) shows that the represented characterizing flavor is, in part or in whole, derived from an artificial flavor as defined by *21 CFR 101.22(a)(1)*.

Q

Does the label or labeling represent the food to be a food that substitutes under 21 CFR 101.13(d) for a standardized food defined in Parts 131 through 169 of 21 CFR and use the name of the standardized food in its statement of identity but does not comply with the standard of identity because of a deviation that is described by an expressed nutrient content claim that has been defined by FDA?

A. If "*YES*," the product must also comply with the requirements of *21 CFR 130.10*.
B. If "*NO*," state that the product is represented as a food provided for under *21 CFR 130.10* but fails to comply with such requirements in that _____ (the blank to be filled with how the product deviates from the requirements).

Illustration VI.A.3
Principal Display Panel
(21 CFR 101.1)

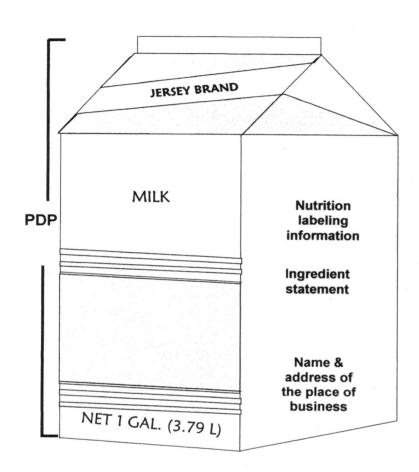

JERSEY BRAND

MILK

PDP

Nutrition labeling information

Ingredient statement

Name & address of the place of business

NET 1 GAL. (3.79 L)

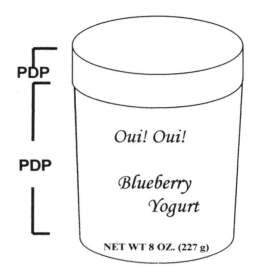

PDP

PDP

Oui! Oui!

Blueberry Yogurt

NET WT 8 OZ. (227 g)

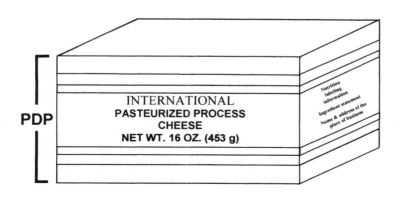

PDP

INTERNATIONAL
PASTEURIZED PROCESS CHEESE
NET WT. 16 OZ. (453 g)

Nutrition labeling information

Ingredient statement

Name & address of the place of business

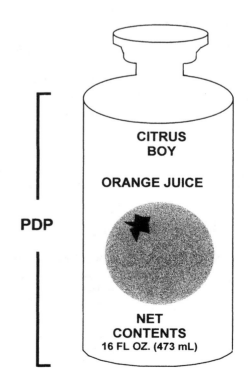

CITRUS BOY

ORANGE JUICE

PDP

NET CONTENTS
16 FL OZ. (473 mL)

Regulation VI.A.4
Foods Named by Use of a Nutrient Content Claim
and a Standardized Term *(21 CFR 130.10)*

Chapter VI

4. Regulation for Foods Named by Use of a Nutrient Content Claim and a Standardized Term *(21 CFR 130.10)*

a) *Description.* The foods prescribed by this general definition and standard of identity are those foods that substitute (see *§101.13(d)* of this chapter) for a standardized food defined in parts 131 through 169 of this chapter and that use the name of that standardized food in their statement of identity but that do not comply with the standard of identity because of a deviation that is described by an expressed nutrient content claim that has been defined by FDA regulation. The nutrient content claim shall comply with the requirements of *§101.13* of this chapter and with the requirements of the regulations in part 101 of this chapter that define the particular nutrient content claim that is used. The food shall comply with the relevant standard in all other respects except as provided in paragraphs (b), (c), and (d) of this section.

b) *Nutrient addition.* Nutrients shall be added to the food to restore nutrient levels so that the product is not nutritionally inferior, as defined in *§101.3(e)(4)* of this chapter, to the standardized food as defined in parts 131 through 169 of this chapter. The addition of nutrients shall be reflected in the ingredient statement.

c) *Performance characteristics.* Deviations from noningredient provisions of the standard of identity (e.g., moisture content, food solids content requirements, or processing conditions) are permitted in order that the substitute food possesses performance characteristics similar to those of the standardized food. Deviations from ingredient and noningredient provisions of the standard must be the minimum necessary to qualify for the nutrient content claim while maintaining similar performance characteristics as the standardized food, or the food will be deemed to be adulterated under section 402(b) of the act. The performance characteristics (e.g., physical properties, flavor characteristics, functional properties, shelf life) of the food shall be similar to those of the standardized food as produced under parts 131 through 169 of this chapter, except that if there is a significant difference in performance characteristics that materially limits the uses of the food compared to the uses of the standardized food, the label shall include a statement informing the consumer of such difference (e.g., if appropriate, *"not recommended for cooking"*). Such statement shall comply with the requirements of *§101.13(d)* of this chapter. The modified product shall perform at least one of the principal functions of the standardized product substantially as well as the standardized product.

d) *Other ingredients.*

 (1) Ingredients used in the product shall be those ingredients provided for by the standard as defined in parts 131 through 169 of this chapter and in paragraph (b) of this section, except that safe and suitable ingredients may be used to improve texture, add flavor, prevent syneresis, extend shelf life, improve appearance, or add sweetness so that the product is not inferior in performance characteristics to the standardized food defined in parts 131 through 169 of this chapter.

 (2) An ingredient or component of an ingredient that is specifically required by the standard (i.e., a mandatory ingredient), as defined in parts 131 through 169 of this chapter, shall not be replaced or exchanged with a similar ingredient from another source unless the standard, as defined in parts 131 through 169 of this chapter, provides for the addition of such ingredient (e.g., vegetable oil shall not replace milkfat in light sour cream).

 (3) An ingredient or component of an ingredient that is specifically prohibited by the standard as defined in parts 131 through 169 of this chapter shall not be added to a substitute food under this section.

 (4) An ingredient that is specifically required by the standard, as defined in parts 131 through 169 of this chapter, shall be present in the product in a significant amount. A significant amount of an ingredient or component of an ingredient is at least that amount that is required to achieve the technical effect of that ingredient in the food.

 (5) Water and fat analogs may be added to replace fat and calories in accordance with *§130.10(c), (d)(1),* and *(d)(2).*

e) *Nomenclature.* The name of a substitute food that complies with all parts of this regulation is the appropriate expressed nutrient content claim and the applicable standardized term.

Adopted from 21 CFR Parts 100 through 169 (4/1/06) Edition)

f) *Label declaration.*

(1) Each of the ingredients used in the food shall be declared on the label as required by the applicable sections of part 101 of this chapter and part 130.

(2) Ingredients not provided for, and ingredients used in excess of those levels provided for, by the standard as defined in parts 131 through 169 of this chapter, shall be identified as such with an asterisk in the ingredient statement, except that ingredients added to restore nutrients to the product as required in paragraph (b) of this section shall not be identified with an asterisk. The statement "*Ingredient(s) not in regular _____" (fill in name of the traditional standardized food) or "*Ingredient(s) in excess of amount permitted in regular _____" (fill in name of the traditional standardized food) or both as appropriate shall immediately follow the ingredient statement in the same type size.

Standards of Identity

Category

1. Milk and Cream *(21 CFR Part 131)*

2. Cheeses and Related Cheese Products (21 *CFR Part 133)*

3. Frozen Desserts *(21 CFR Part 135)*

4. Bakery Products *(21 CFR Part 136)*

5. Cereal Flours and Related Products *(21 CFR Part 137)*

6. Macaroni and Noodle Products *(21 CFR Part 139)*

7. Canned Fruits *(21 CFR Part 145)*

8. Canned Fruit Juices *(21 CFR Part 146)*

9. Fruit Butters, Jellies, Preserves, and Related Products *(21 CFR Part 150)*

10. Canned Vegetables *(21 CFR Part 155)*

11. Vegetable Juices *(21 CFR Part 156)*

12. Frozen Vegetables *(21 CFR Part 158)*

13. Eggs and Egg Products *(21 CFR Part 160)*

14. Fish and Shellfish *(21 CFR Part 161)*

15. Cacao Products *(21 CFR Part 163)*

16. Tree Nut and Peanut Products *(21 CFR Part 164)*

17. Beverages *(21 CFR Part 165)*

18. Margarine *(21 CFR Part 166)*

19. Sweeteners and Table Sirups *(21 CFR Part 168)*

20. Food Dressings and Flavorings *(21 CFR Part 169)*

Adopted from 21 CFR Parts 100 through 169 (4/1/06 Edition)

1. MILK AND CREAM *(21 CFR Part 131)*

Subpart A - General Provisions

Section	
131.3	Definitions
131.25	Whipped cream products containing flavoring or sweetening

Subpart B - Requirements for Specific Standardized Milk and Cream

Section	
131.110	Milk
131.111	Acidified milk
131.112	Cultured milk
131.115	Concentrated milk
131.120	Sweetened condensed milk
131.125	Nonfat dry milk
131.127	Nonfat dry milk fortified with vitamins A and D
131.130	Evaporated milk
131.147	Dry whole milk
131.149	Dry cream
131.150	Heavy cream
131.155	Light cream
131.157	Light whipping cream
131.160	Sour cream
131.162	Acidified sour cream
131.170	Eggnog
131.180	Half-and-half
131.200	Yogurt
131.203	Lowfat yogurt
131.206	Nonfat yogurt

2. CHEESES AND RELATED CHEESE PRODUCTS *(21 CFR Part 133)*

Subpart A - General Provisions

Section	
133.3	Definitions
133.5	Methods of analysis
133.10	Notice to manufacturers, packers, and distributors of pasteurized blended cheese, pasteurized process cheese, cheese food, cheese spread, and related foods

Subpart B - Requirements for Specific Standardized Cheese and Related Products

Section	
133.102	Asiago fresh and asiago soft cheese
133.103	Asiago medium cheese
133.104	Asiago old cheese
133.106	Blue cheese
133.108	Brick cheese
133.109	Brick cheese for manufacturing
133.111	Caciocavallo siciliano cheese
133.113	Cheddar cheese
133.114	Cheddar cheese for manufacturing
133.116	Low sodium cheddar cheese
133.118	Colby cheese
133.119	Colby cheese for manufacturing
133.121	Low sodium colby cheese
133.123	Cold-pack and club cheese
133.124	Cold-pack cheese food
133.125	Cold-pack cheese food with fruits, vegetables, or meats
133.127	Cook cheese, koch kaese
133.128	Cottage cheese
133.129	Dry curd cottage cheese
133.133	Cream cheese
133.134	Cream cheese with other foods
133.136	Washed curd and soaked curd cheese
133.137	Washed curd cheese for manufacturing
133.138	Edam cheese
133.140	Gammelost cheese
133.141	Gorgonzola cheese
133.142	Gouda cheese
133.144	Granular and stirred curd cheese
133.145	Granular cheese for manufacturing
133.146	Grated cheeses
133.147	Grated American cheese food
133.148	Hard grating cheeses
133.149	Gruyere cheese
133.150	Hard cheeses
133.152	Limburger cheese
133.153	Monterey cheese and monterey jack cheese
133.154	High-moisture jack cheese
133.155	Mozzarella cheese and scamorza cheese

Chart VI.A.6
(Continued)

Chapter VI

Subpart B: Standardized Cheese (continued)

133.156	Low-moisture mozzarella and scamorza cheese
133.157	Part-skim mozzarella and scamorza cheese
133.158	Low-moisture part-skim mozzarella and scamorza cheese
133.160	Muenster and munster cheese
133.161	Muenster and munster cheese for manufacturing
133.162	Neufchatel cheese
133.164	Nuworld cheese
133.165	Parmesan and reggiano cheese
133.167	Pasteurized blended cheese
133.168	Pasteurized blended cheese with fruits, vegetables, or meats
133.169	Pasteurized process cheese
133.170	Pasteurized process cheese with fruits, vegetables, or meats
133.171	Pasteurized process pimento cheese
133.173	Pasteurized process cheese food
133.174	Pasteurized process cheese food with fruits, vegetables, or meats
133.175	Pasteurized cheese spread
133.176	Pasteurized cheese spread with fruits, vegetables, or meats
133.178	Pasteurized neufchatel cheese spread with other foods
133.179	Pasteurized process cheese spread
133.180	Pasteurized process cheese spread with fruits, vegetables, or meats
133.181	Provolone cheese
133.182	Soft ripened cheeses
133.183	Romano cheese
133.184	Roquefort cheese, sheep's milk blue-mold, and blue-mold cheese from sheep's milk
133.185	Samsoe cheese
133.186	Sap sago cheese
133.187	Semisoft cheeses
133.188	Semisoft part-skim cheeses
133.189	Skim milk cheese for manufacturing
133.190	Spiced cheeses
133.191	Part-skim spiced cheeses
133.193	Spiced, flavored standardized cheeses
133.195	Swiss and emmentaler cheese
133.196	Swiss cheese for manufacturing

3. FROZEN DESSERTS *(21 CFR Part 135)*

Subpart A - General Provisions

Section

135.3	Definitions

Subpart B - Requirements for Specific Standardized Frozen Desserts

Section

135.110	Ice cream and frozen custard
135.115	Goat's milk ice cream
135.130	Mellorine
135.140	Sherbet
135.160	Water ices

4. BAKERY PRODUCTS *(21 CFR Part 136)*

Subpart A - General Provisions

Section

136.3	Definitions

Subpart B - Requirements for Specific Standardized Bakery Products

Section

136.110	Bread, rolls, and buns
136.115	Enriched bread, rolls, and buns
136.130	Milk bread, rolls, and buns
136.160	Raisin bread, rolls, and buns
136.180	Whole wheat bread, rolls, and buns

5. CEREAL FLOURS AND RELATED PRODUCTS *(21 CFR Part 137)*

Subpart A - [Reserved]

Subpart B - Requirements for Specific Standardized Cereal Flours and Related Products

Section

137.105	Flour
137.155	Bromated flour
137.160	Enriched bromated flour
137.165	Enriched flour
137.170	Instantized flours
137.175	Phosphated flour

Adopted from 21 CFR Parts 100 through 169 (4/1/06 Edition)

Subpart B – Standardized Cereal Flour
(continued)

137.180	Self-rising flour
137.185	Enriched self-rising flour
137.190	Cracked wheat
137.195	Crushed wheat
137.200	Whole wheat flour
137.205	Bromated whole wheat flour
137.211	White corn flour
137.215	Yellow corn flour
137.220	Durum flour
137.225	Whole durum flour
137.250	White corn meal
137.255	Bolted white corn meal
137.260	Enriched corn meals
137.265	Degerminated white corn meal
137.270	Self-rising white corn meal
137.275	Yellow corn meal
137.280	Bolted yellow corn meal
137.285	Degerminated yellow corn meal
137.290	Self-rising yellow corn meal
137.300	Farina
137.305	Enriched farina
137.320	Semolina
137.350	Enriched rice

6. MACARONI AND NOODLE PRODUCTS
(21 CFR Part 139)

Subpart A - [Reserved]

Subpart B - Requirements for Specific Standardized Macaroni and Noodle Products

Section

139.110	Macaroni products
139.115	Enriched macaroni products
139.117	Enriched macaroni products with fortified protein
139.120	Milk macaroni products
139.121	Nonfat milk macaroni products
139.122	Enriched nonfat milk macaroni products
139.125	Vegetable macaroni products
139.135	Enriched vegetable macaroni products
139.138	Whole wheat macaroni products
139.140	Wheat and soy macaroni products
139.150	Noodle products
139.155	Enriched noodle products
139.160	Vegetable noodle products
139.165	Enriched vegetable noodle products
139.180	Wheat and soy noodle products

7. CANNED FRUITS *(21 CFR Part 145)*

Subpart A - General Provisions

Section

145.3	Definitions

Subpart B - Requirements for Specific Standardized Canned Fruits

Section

145.110	Canned applesauce
145.115	Canned apricots
145.116	Artificially sweetened canned apricots
145.120	Canned berries
145.125	Canned cherries
145.126	Artificially sweetened canned cherries
145.130	Canned figs
145.131	Artificially sweetened canned figs
145.134	Canned preserved figs
145.135	Canned fruit cocktail
145.136	Artificially sweetened canned fruit cocktail
145.140	Canned seedless grapes
145.145	Canned grapefruit
145.170	Canned peaches
145.171	Artificially sweetened canned peaches
145.175	Canned pears
145.176	Artificially sweetened canned pears
145.180	Canned pineapple
145.181	Artificially sweetened canned pineapple
145.185	Canned plums
145.190	Canned prunes

8. CANNED FRUIT JUICES *(21 CFR Part 146)*

Subpart A - General Provisions

Section

146.3	Definitions

Chart VI.A.6

(Continued)

Chapter VI

Subpart B – Requirements for Specific Standardized Canned Fruit Juices and Beverages

Section
146.114	Lemon juice
146.120	Frozen concentrate for lemonade
146.121	Frozen concentrate for artificially sweetened lemonade
146.126	Frozen concentrate for colored lemonade
146.132	Grapefruit juice
146.135	Orange juice
146.137	Frozen orange juice
146.140	Pasteurized orange juice
146.141	Canned orange juice
146.145	Orange juice from concentrate
146.146	Frozen concentrated orange juice
146.148	Reduced acid frozen concentrated orange juice
146.150	Canned concentrated orange juice
146.151	Orange juice for manufacturing
146.152	Orange juice with preservative
146.153	Concentrated orange juice for manufacturing
146.154	Concentrated orange juice with preservative
146.185	Pineapple juice
146.187	Canned prune juice

9. FRUIT BUTTERS, JELLIES, PRESERVES, AND RELATED PRODUCTS *(21 CFR Part 150)*

Subpart A - [Reserved]

Subpart B - Requirements for Specific Standardized Fruit Butters, Jellies, Preserves, and Related Products

Section
150.110	Fruit butter
150.140	Fruit jelly
150.141	Artificially sweetened fruit jelly
150.160	Fruit preserves and jams
150.161	Artificially sweetened fruit preserves and jams

10. CANNED VEGETABLES *(21 CFR Part 155)*

Subpart A - General Provisions

Section
155.3	Definitions

Subpart B - Requirement for Specific Standardized Canned Vegetables

Section
155.120	Canned green beans and canned wax beans
155.130	Canned corn
155.131	Canned field corn
155.170	Canned peas
155.172	Canned dry peas
155.190	Canned tomatoes
155.191	Tomato concentrates
155.194	Catsup
155.200	Certain other canned vegetables
155.201	Canned mushrooms

11. VEGETABLE JUICES *(21 CFR Part 156)*

Subpart A - General Provisions

Section
156.3	Definitions

Subpart B - Requirements for Specific Standardized Vegetable Juices

Section
156.145	Tomato juice

12. FROZEN VEGETABLES *(21 CFR Part 158)*

Subpart A - General Provisions

Section
158.3	Definitions

Subpart B - Requirements for Specific Standardized Frozen Vegetables

Section
158.170	Frozen peas

Adopted from 21 CFR Parts 100 through 169 (4/1/06 Edition)

Chart VI.A.6
(Continued)

13. EGGS AND EGG PRODUCTS *(21 CFR Part 160)*

Subpart A - [Reserved]

Subpart B - Requirements for Specific Standardized Eggs and Egg Products

Section
160.100   Eggs
160.105   Dried eggs
160.110   Frozen eggs
160.115   Liquid eggs
160.140   Egg whites
160.145   Dried egg whites
160.150   Frozen egg whites
160.180   Egg yolks
160.185   Dried egg yolks
160.190   Frozen egg yolks

14. FISH AND SHELLFISH *(21 CFR Part 161)*

Subpart A - General Provisions

Section
161.30   Declaration of quantity of contents on labels for canned oysters

Subpart B - Requirements for Specific Standardized Fish and Shellfish

Section
161.130   Oysters
161.136   Olympia oysters
161.145   Canned oysters
161.170   Canned Pacific salmon
161.173   Canned wet pack shrimp in transparent or nontransparent containers
161.175   Frozen raw breaded shrimp
161.176   Frozen raw lightly breaded shrimp
161.190   Canned tuna

15. CACAO PRODUCTS *(21 CFR Part 163)*

Subpart A - General Provisions

Section
163.5   Methods of analysis

Subpart B - Requirements for Specific Standardized Cacao Products

Section
163.110   Cacao nibs
163.111   Chocolate liquor
163.112   Breakfast cocoa
163.113   Cocoa
163.114   Lowfat cocoa
163.117   Cocoa with dioctyl sodium sulfosuccinate for manufacturing
163.124   White chocolate
163.123   Sweet chocolate
163.130   Milk chocolate
163.135   Buttermilk chocolate
163.140   Skim milk chocolate
163.145   Mixed dairy product chocolates
163.150   Sweet cocoa and vegetable fat coating
163.153   Sweet chocolate and vegetable fat coating
163.155   Milk chocolate and vegetable fat coating

16. TREE NUT AND PEANUT PRODUCTS *(21 CFR Part 164)*

Subpart A - [Reserved]

Subpart B - Requirements for Specific Standardized Tree Nut and Peanut Products

Section
164.110   Mixed nuts
164.120   Shelled nuts in rigid or semirigid containers
164.150   Peanut butter

17. BEVERAGES *(21 CFR Part 165)*

Subpart A - General Provisions

Section
165.3   Definitions

Subpart B - Requirements for Specific Standardized Beverages

165.110   Bottled water

Chart VI.A.6	Chapter VI
(Continued)	

18. MARGARINE *(21 CFR Part 166)*

Subpart A - General Provisions

Section
166.40   Labeling of margarine

Subpart B - Requirements for Specific Standardized Margarine

Section
168.110   Margarine

19. SWEETENERS AND TABLE SIRUPS *(21 CFR Part 168)*

Subpart A - [Reserved]

Subpart B - Requirements for Specific Standardized Sweeteners and Table Sirups

Section
168.110   Dextrose anhydrous
168.111   Dextrose monohydrate
168.120   Glucose sirup
168.121   Dried glucose sirup
168.122   Lactose
168.130   Cane sirup

168.140   Maple sirup
168.160   Sorghum sirup
168.180   Table sirup

20. FOOD DRESSINGS AND FLAVORINGS *(21 CFR Part 169)*

Subpart A - General Provisions

Section
169.3   Definitions

Subpart B - Requirements for Specific Standardized Food Dressings and Flavorings

Section
169.115   French dressing
169.140   Mayonnaise
169.150   Salad dressing
169.175   Vanilla extract
169.176   Concentrated vanilla extract
169.177   Vanilla flavoring
169.178   Concentrated vanilla flavoring
169.179   Vanilla powder
169.180   Vanilla-vanillin extract
169.181   Vanilla-vanillin flavoring
169.182   Vanilla-vanillin powder

Adopted from 21 CFR Parts 100 through 169 (4/1/06 Edition)

Subpart A - General Provisions

Section

102.5	General principles
102.19	Petitions

Subpart B - Requirements For Specific Nonstandardized Foods

Section

102.22	Protein hydrolysates
102.23	Peanut spreads
102.26	Frozen *"heat and serve"* dinners
102.28	Foods packaged for use in the preparation of *"main dishes"* or *"dinners"*

102.33	Beverages that contain fruit or vegetable juice
102.37	Mixtures of edible fat or oil and olive oil
102.39	Onion rings made from diced onion
102.41	Potato chips made from dried potatoes
102.45	Fish sticks or portions made from minced fish
102.46	Pacific whiting
102.47	Bonito
102.49	Fried clams made from minced clams
102.50	Crabmeat
102.54	Seafood cocktails
102.55	Nonstandardized breaded composite shrimp units
102.57	Greenland turbot *(Reinhardtius hippoglossoides)*

a) The principal display panel of a food in packaged form shall bear as one of its principal features, a statement of the identity of the commodity.
b) Such statement of identity shall be in terms of:
   (1) The name now or hereafter specified in or required by any applicable Federal law or regulation; or, in the absence thereof,
   (2) The common or usual name of the food, or, in the absence thereof,
   (3) An appropriately descriptive term, or when the nature of the food is obvious, a fanciful name commonly used by the public for such food.
c) Where a food is marketed in various optional forms (whole, slices, diced, etc.), the particular form shall be considered to be a necessary part of the statement of identity and shall be declared in letters of a type size bearing a reasonable relation to the size of the letters forming the other components of the statement of identity; except that if the optional form is visible through the container or is depicted by an appropriate vignette, the particular form need not be included in the statement. This specification does not affect the required declarations of identity under definitions and standards for foods promulgated pursuant to section 401 of the act.
d) This statement of identity shall be presented in bold type on the principal display panel, shall be in a size reasonably related to the most prominent printed matter on such panel, and shall be in lines generally parallel to the base on which the package rests as it is designed to be displayed.
e) Under the provisions of section 403(c) of the Federal Food, Drug, and Cosmetic Act, a food shall be deemed to be misbranded if it is an imitation of another food unless its label bears, in type of uniform size and prominence the word "imitation" and, immediately thereafter, the name of the food imitated.
   (1) A food shall be deemed to be an imitation and thus subject to the requirements of section 403(c) of the act if it is a substitute for and resembles another food but is nutritionally inferior to that food.
   (2) A food that is a substitute for and resembles another food shall not be deemed to be an imitation provided it meets each of the following requirements:
      (i) It is not nutritionally inferior to the food for which it substitutes and which it resembles.
      (ii) Its label bears a common or usual name that complies with the provisions of §102.5 of this chapter and that is not false or misleading, or in the absence of an existing common or usual name, an appropriately descriptive term that is not false or misleading. The label may, in addition, bear a fanciful name which is not false or misleading.
   (3) A food for which a common or usual name is established by regulation (e.g., in a standard of identity pursuant to section 401 of the act, in a common or usual name regulation pursuant to part 102 of this chapter, or in a regulation establishing a nutritional quality guideline pursuant to part 104 of this chapter), and which complies with all of the applicable requirements of such regulation(s), shall not be deemed to be an imitation.
   (4) Nutritional inferiority includes:
      (i) Any reduction in the content of an essential nutrient that is present in a measurable amount, but does not include a reduction in the caloric or fat content provided the food is labeled pursuant to the provisions of §101.9, and provided the labeling with respect to any reduction in caloric content complies with the provisions applicable to caloric content in part 105 of this chapter.
      (ii) For the purpose of this section, a measurable amount of an essential nutrient in a food shall be considered to be 2 percent or more of the Daily Reference Value (DRV) of protein listed under §101.9(c)(7)(iii) and of potassium listed under §10l.9(c)(9) per reference amount customarily consumed and 2 percent or more of the Reference Daily Intake (RDI) of any vitamin or mineral listed under §101.9(c)(8)(iv) per reference amount customarily consumed, except that selenium, molybdenum, chromium, and chloride need not be considered.
      (iii) If the Commissioner concludes that a food is a substitute for and resembles another food but is inferior to the food imitated for reasons other than those set forth in this paragraph, he may propose appropriate revisions to this regulation or he may propose a separate regulation governing the particular food.

f) A label may be required to bear the percentage(s) of a characterizing ingredient(s) or information concerning the presence or absence of an ingredient(s) or the need to add an ingredient(s) as part of the common or usual name of the food pursuant to subpart B of part 102 of this chapter.

A food shall be deemed to be an imitation and thus subject to the requirements of section 403(c) of the act if it is a substitute for and resembles another food and is *nutritionally inferior* to that food. *(21 CFR 101.3(e)(1))*

Food Status	Requirements
Nutritionally Inferior	(i) Any reduction in the content of an essential nutrient that is present in a measurable amount (2% or more of the DRV and/or RDI nutrients per reference amount customarily consumed) in the traditional food, except for a reduction in the caloric or fat content provided that the food is labeled pursuant to the provisions of section 101.9, and provided the labeling with respect to any reduction in caloric content complies with the provisions applicable to caloric content in *21 CFR Part 105*. The content of selenium, molybdenum, chromium, and chloride need not be considered for nutritional inferiority. (ii) Shall bear the word "imitation" and immediately thereafter, the name of the food imitated in type of uniform size and prominence as the name of the traditional food imitated.
Not Nutritionally Inferior	(i) Shall not be deemed to be an imitation. (ii) Is not required to bear the word "imitation" on its label or labeling. (iii) May use sufficiently descriptive terms or phrases that are not false and/or misleading with such terms as "substitute", "replacer", or "alternative", etc.

Adopted from 21 CFR Parts 100 through 169 (4/1/06 Edition)

Notes:

1. **Definitions:**

   a) Artificial Flavor - any substance, the function of which is to impart flavor, which is not derived from a spice, fruit or fruit juice, vegetable or vegetable juice, edible yeast, herb, bark, bud, root, leaf or similar plant material, meat, fish, poultry, eggs, dairy products, or fermentation products thereof. Artificial flavor includes the substances listed in §§ 172.515(b) and 182.60, except when these are derived from natural sources.

   b) Spice – any aromatic vegetable substance in whole, broken, or ground form, except those substances which have been regarded as foods, such as onions, garlic, and celery; whose significant function in food is seasoning rather than nutritional; that is true to name; and from which no portion of any volatile oil or other flavoring principle has been removed. (For spice names, see §§ 101.22(a)(2) and 182.10, and *Part 184.*)

   Paprika, turmeric, and saffron or other spices which are also colors, shall be declared as "spice and coloring" unless declared by their common or usual names.

   c) Natural Flavor – essential oil, oleoresin, essence or extractive, protein hydrolysate, distillate, or any product of roasting, heating, or enzymolysis, which contains the flavoring constituents derived from a spice, fruit or fruit juice, vegetable or vegetable juice, edible yeast, herb, bark, bud, root, leaf or similar plant material, meat, seafood, poultry, eggs, dairy products, or fermentation products thereof, whose significant function in food is flavoring rather than nutritional. Natural flavors include the natural essence or extractives obtained from plants listed in §§182.10, 182.20, 182.40, and 182.50 and *Part 184* of this chapter, and the substances listed in § 172.510 of this chapter.

Chart VI.A.10
(Continued)

Chapter VI

2. FLAVORS SHIPPED TO MANUFACTURERS OR PROCESSORS

A flavor shipped to manufacturers or processors (but not to consumers) for use in the manufacture of a fabricated food shall be labeled in one of the following ways, unless it is one for which a standard of identity has been promulgated, in which case it shall be labeled as provided in the standard. *(21 CFR 101.22(g))*

Nature of the Flavor	Comments on How the Flavor is to be Labeled	CFR Reference
1. Flavor consists of one ingredient.	Declare flavor by its common or usual name.	*§101.22(g)(1)*
2. Flavor consists of two or more ingredients.	a) Declare each ingredient in the flavor by its common or usual name; or b) State the following: *"All flavor ingredients contained in this product are approved for use in a regulation of the Food and Drug Administration."* Any flavor ingredient not contained in one of these regulations, and any nonflavor ingredients, shall be listed separately on the label.	*§101.22(g)(2)*
3. Flavor consists solely of natural flavor(s).	Flavor shall be so labeled, e.g., *"Strawberry Flavor," "Banana Flavor,"* or *"Natural Strawberry Flavor."*	*§101.22(g)(3)*
4. Flavor consists of both natural flavor and artificial flavor.	Flavor shall be so labeled, e.g., *"Natural and Artificial Strawberry Flavor."*	*§101.22(g)(3)*
5. Flavor consists solely of artificial flavor.	Flavor shall be so labeled, e.g., *"Artificial Strawberry Flavor."*	*§101.22(g)(3)*

Adopted from 21 CFR Parts 100 through 169 (4/1/06 Edition)

3. FLAVORS USED BY CONSUMERS OR IN CONSUMER PRODUCTS

PART 1. Ingredient Labeling

The label of a food to which a flavor is added shall declare the flavor in the statement of ingredients in the following ways: *(21CFR 101.22(h))*

Nature of the Flavor or Flavoring Ingredient	Requirements for Labeling of Flavor or Flavoring Ingredient	CFR Reference
1. Spice, natural flavor, and artificial flavor.	May be declared as *"spice," "natural flavor," and "artificial flavor,"* or any combination thereof, as the case may be.	*§101.22(h)(1)*
2. Incidental additives originating in a flavor or spice used in the manufacture of a food.	Need not be declared in statement of ingredients if it meets the requirements of § *101.100(a)(3)*.	*§101.22(h)(2)*
3. Substances obtained by cutting, grinding, drying, pulping, or similar processing of tissues derived from fruit, vegetable, meat, fish, or poultry that are used for flavoring.	Shall be declared as the food by its common or usual name, e.g., powdered or granulated onions, garlic powder, or celery powder rather than as a flavor.	*§101.22(h)(3)*

PART 2. Characterizing Flavor Declarations Other Than In The Ingredient Statement

When the label, labeling, or advertising of a food makes any direct or indirect representation with respect to a primary recognizable flavor(s), by word, vignette, e.g., depiction of a fruit, or other means, or if for any other reason the manufacturer or distributor of a food wishes to designate the type of flavor in the food other than through the statement of ingredients, such flavor shall be considered to be the *"characterizing flavor"* and shall be declared in the following ways: *(21 CFR 101.22(i)(1))*

Nature of the Characterizing Flavor or Flavoring Ingredients	Requirements for Labeling of the Characterizing Flavor or Flavoring Ingredients	CFR Reference
1. Food contains no artificial flavor, which simulates, resembles, or reinforces the characterizing flavor.	Name of the food on the principal display panel shall be accompanied by the common or usual name or the characterizing flavor, e.g., *"vanilla,"* in letters not less than one-half the height of the letters used in the name of the food except that:(see 1a, b, and c below)	*§101.22(i)(1)*

Chart VI.A.10
(Continued)

Chapter VI

Nature of the Characterizing Flavor or Flavoring Ingredients	Requirements for Labeling of the Characterizing Flavor or Flavoring Ingredients	CFR Reference
a) If the food is one that is expected to contain a characterizing ingredient, e.g., *"strawberries in strawberry shortcake,"* and the food contains natural flavor derived from the characterizing ingredient and in an amount insufficient to independently characterize the food, or the food contains no such ingredient.	The name of the characterizing flavor may be immediately preceded by the word *"natural"* and shall be immediately followed by the word *"flavored"* in letters not less than one-half the height of the letters used in the name of the characterizing flavor, e.g., *"natural strawberry flavored shortcake"* or *"strawberry flavored shortcake."*	*§101.22(i)(1)(i)*
b) If none of the natural flavor used in the food is derived from the product whose flavor is simulated, e.g., lime used as flavor in a product labeled as lemon drink.	The product is labeled either with the flavor of the product from which the flavor is derived, e.g., *"lime flavor"*, or as *"artificially lemon flavored"*.	*§101.22(i)(1)(ii)*
c) If the food contains both a characterizing flavor from the product whose flavor is simulated or other natural flavor which simulates, resembles, or reinforces the characterizing flavor.	The product is labeled with the name of the characterizing flavor immediately preceded by the word *"natural"* and shall be immediately followed by the word *"flavored"* and the name of the food shall be immediately followed by the words *"with other natural flavors"* in letters not less than one-half the height of the letters in the name of the characterizing flavor.	*§101.22(i)(1)(iii)*

Adopted from 21 CFR Parts 100 through 169 (4/1/06 Edition)

Chart VI.A.10
(Continued)

Nature of the Characterizing Flavor or Flavoring Ingredients	Requirements for Labeling of the Characterizing Flavor or Flavoring Ingredients	CFR Reference
2. Food contains artificial flavor which simulates, resembles, or reinforces the characterizing flavor.	Name of the food on the principal display panel or other panels of the label shall be accompanied by the common or usual name(s) of the characterizing flavor, in letters not less than one-half the height of the letters used in the name of the food; and the name of the characterizing flavor shall be accompanied by the word(s) *"artificial"* or *"artificially flavored,"* in letters not less than one-half the height of the letters in the name of the characterizing flavor, e.g., *"artificial vanilla,"* or *"artificially flavored strawberry,"* or *"grape artificially flavored."*	*§101.22(i)(2)*

4. FLAVOR LABELING EXCEPTIONS AND VARIATIONS *(21 CFR 101.22(i)(3))*

Wherever the name of the characterizing flavor appears on the label (other than in the ingredient statement) so conspicuously as to be easily seen under customary conditions of purchase, the words prescribed by *§101.22(i)* must immediately precede or follow such name, without any intervening written, printed, or graphic matter, except:

1. Where the characterizing flavor and a brand or trademark are presented together.	Other written, printed, or graphic matter that is a part of or is associated with the trademark or brand may intervene if the required words are in such relationship with the trademark or brand as to be clearly related to the characterizing flavor.	*§101.22(i)(3)(i)*
2. If the food contains more than one flavor subject to these regulations.	Statements required by *§101.22,* e.g., *"artificially flavored,"* need appear only once in each statement of characterizing flavors present in such food, e.g., *"artificially flavored vanilla and strawberry."*	*§101.22(i)(3)(ii)*
3. If the food contains three or more distinguishable characterizing flavors, or a blend of flavors with no primary recognizable flavor.	Flavor may be declared by an appropriately descriptive generic term in lieu of naming each flavor, e.g., *"artificially flavored fruit punch."*	*§101.22(i)(3)(iii)*

Adopted from 21 CFR Parts 100 through 169 (4/1/06 Edition)

Chart VI.A.10
(Continued)

Chapter VI

5. FLAVOR CERTIFICATION *(21 CFR 101.22(i)(4)(i) through (v))*

A flavor supplier shall certify, in writing, that any flavor he supplies which is designated as containing no artificial flavor does not, to the best of his knowledge and belief, contain any artificial flavor, and that he has added no artificial flavor to it. The requirement for such certification may be satisfied by a guarantee under section 303(c)(2) of the FD&CA which contains such a specific statement. A flavor user shall be required to make such a written certification only where he adds to or combines another flavor with a flavor which has been certified by a flavor supplier as containing no artificial flavor, but otherwise such user may rely upon the supplier's certification and need make no separate certification.

All such certifications shall be retained by the certifying party throughout the period in which the flavor is supplied and for a minimum of three years thereafter, and shall be subject to the following conditions:

(1) The certifying party shall make such certifications available upon request at all reasonable hours to any duly authorized office or employee of the Food and Drug Administration or any other employee acting on behalf of the Secretary of Health and Human Services. *(21 CFR 101.22(i)(4)(i))*

(2) Wherever possible the Food and Drug Administration shall verify the accuracy of a reasonable number of certifications made pursuant to this section. *(21 CFR 101.22(i)(4)(ii))*

(3) Where no person authorized to provide such information is reasonably available at the time of inspection, the certifying party shall arrange to have such person and the relevant material and records ready for verification as soon as practicable: Provided, that whenever the Food and Drug Administration has reason to believe that the supplier or user may utilize this period to alter inventories or records, such additional time shall not be permitted.

Where such additional time is provided FDA may require the certifying party to certify that relevant inventories have not been materially disturbed and that relevant records have not been altered or concealed during such period. *(21 CFR 101.22(i)(4)(iii))*

(4) The certifying party shall provide, to any officer or representative duly designated by the Secretary of DHHS, such qualitative statement of the composition of the flavor or product covered by the certification as may be reasonably expected to enable the Secretary's representatives to determine which relevant raw and finished materials and flavor ingredient records are reasonably necessary to verify the certifications. The examination conducted by the Secretary's representative shall be limited to inspection and review of inventories and ingredient records for those certifications which are to be verified. *(21 CFR 101.22(i)(4)(iv))*

(5) Review of flavor ingredient records shall be limited to the quantitative formula. The person verifying the certifications may only make such notes as are necessary to enable him to verify the certification. *(21 CFR 101.22(i)(4)(v))*

Chapter VI

Section B: Designation of Ingredients *(21 CFR 101.4)*

1. **Summary of the Requirements**

   a) Multicomponent foods in package form shall bear an ingredient statement that lists all ingredients, except when exempted by *21 CFR 101.100.* (See Regulations IX.B)

   b) Ingredients that are required to be declared on the label of a food shall be declared by their common or usual names and not by a collective (generic) term or name, except that:

      (1) Spices, flavorings, colorings, and chemical preservatives shall be declared as required by *21 CFR 101.22,* and

      (2) An ingredient which itself contains two or more ingredients and has a common or usual name may be declared in accordance with *§101.4(b)(2)* by:

         (i) Its common or usual name followed by a parenthetical listing of all ingredients contained therein in descending order of predominance, or

         (ii) Declaring each ingredient in the multicomponent food by its common or usual name in descending order of predominance in the finished food without declaring the multicomponent ingredient itself.

      (3) The foods covered by *§101.4(b)(3) through (22)* may be declared as specified by the specific regulation. (See Regulations VI.B.4)

   c) Ingredients that are required to be declared on the label shall be listed on the principal display panel; or if the label has an information panel, they shall be listed on the information panel together with the nutrition information, name and address of the place of business without intervening matter, except that under *21 CFR 101.2(d)(2),* if the package consists of a separate lid and body, bears nutrition labeling and the lid qualifies for and is designated as the principal display panel:

      (1) The name and address of the business information required by *21 CFR 101.5* shall not be required on the body of the container if the information appears on the lid;

      (2) The nutrition information required by *21 CFR 101.9* shall not be required on the lid if the information appears on the body of the container; and

      (3) The statement of ingredients required by *§101.4* shall not be required on the lid if the information appears on the body of the container. In addition, the statement of ingredients is not required on the container body if this information appears on the lid in accordance with *§101.2.*

   d) Ingredients that are required to be declared shall be listed in descending order of predominance by weight, except:

      (1) Ingredients in amounts of 2 percent or less may be placed at the end of the listing followed by an appropriate quantifying statement (*§101.4(a)(2));*

      (2) When water is added to completely or partially reconstitute an ingredient, permitted under *§101.4(b)(3) through (12)* to be declared by its class name, to single strength. The position of the ingredient class name in the ingredient statement shall be determined by the weight of the unreconstituted ingredient plus the weight of the water added to reconstitute the ingredient to single strength. Any water added in excess of that amount shall be declared in the ingredient statement as water *(21 CFR 101.4(c));*

      (3) When a food that is characterized on the label as "nondairy" contains a caseinate ingredient, the caseinate ingredient shall be followed by a parenthetical statement identifying its source, e.g., sodium caseinate (a milk derivative), in the statement of ingredients *(§101.4(d));*

      (4) When the percentage of an ingredient is included in the statement of ingredients, it shall be shown in parentheses following the name of the ingredient to the nearest 1 percent *(21 CFR 101.4(e));* and

      (5) A sulfiting agent (e.g., sodium sulfite) that has been added to any food or to any ingredient in a food, and that has no technical effect in that food, will be considered to be present in an insignificant amount only if no detectable amount of the agent is present in the finished food.

   e) The statement of ingredients shall be prominently and conspicuously declared, but in no case may the letters and/or numbers be less than one-sixteenth inch in height unless:

The label of the package is too small to accommodate all of the required information. In such case, FDA may establish by regulation an acceptable alternative method of disseminating such information to the public, e.g., labeling attached to or inserted in a package or available at the point of purchase, or may establish a type size smaller than one-sixteenth inch in height;

(f) The ingredient statement must declare major allergens when present in a food, including flavors and colors, by:

    (1) Declaring the major allergen's name in an additional statement with "contains" immediately after or adjacent to the ingredient list in a type size at least the same as the ingredient list, or

    (2) Add the major food allergen's name in parenthesis after the name of the ingredient in the ingredient statement.

2. Questions and Responses

The questions and responses sections included under Chapter VI are a series of established requirements designed to aid the reviewers in establishing the degree of compliance that a specific food label and its labeling comply with applicable laws and regulations. It is not intended to represent all questions that may arise during the review process, but as examples of typical questions during a label review. This section can also serve as a teaching aid for less experienced reviewers and as a ready reference for experienced reviewers. It can also serve as an example for developing a response to advise those responsible for food labels and labeling of needed corrections.

Is the food fabricated from two or more ingredients?

A. If "*YES*," continue.

B. If "*NO*," STOP HERE, an ingredient statement is not required.

Does the label bear an ingredient statement?

A. If "*YES*," continue.

B. If "*NO*," state that the product is fabricated from two or more ingredients, but its label fails to declare each ingredient in the food.

Does the label bear an information panel? *(See Illustration VI.B.3)*

A. If "*YES*," the statement of ingredients shall appear on such panel together with the name and address of the responsible firm and the nutrition information without intervening matter, unless exempted under *21 CFR 101.100* (See Exemption IX.B), *and 21 CFR 101.2(d).* or exempted in accordance with *21 CFR 101.9(j).* (See Regulations VI.E.15)

B. If "*NO*," because the label does not bear an information panel, the ingredient statement and other required labeling information shall appear on the principal display panel.

If "*NO*," because it is not declared on the information panel and is not exempt as provided for above, state that label bears an information panel, but _____ (insert the specific label component that does not appear) fails to appear on such panel, together with the _____, _____, and _____(insert components that appear on the panel).

Does the statement of ingredients appear prominently and conspicuously on the label (the letters and numbers may not be less than one-sixteenth inch in height, unless exempted)? *(21 CFR 101.2(c))*

A. If "*YES*," continue.

B. If "*NO*," state that the statement of ingredients fails to appear prominently and conspicuously on the label due to _____ (insert the specific reason, e.g., its not being declared in the required type size, a lack of sufficient color contrast between the declaration and the background, etc.).

 Is each ingredient declared by its specific common or usual name and not a collective or generic name as provided for in 21 CFR 101.4(a)(1), except for ingredients exempted by 21 CFR 101.100, 101.4(b)(3) through (22), and 101.4(c)? (See Regulations VI.B.4 and IX.E)

A. If "*YES*," it may bear a claim, e.g., low sodium potato chips, continue.
B. If "*NO*," because the product bears a claim for which it qualifies, but has not been specially processed, altered, formulated, or reformulated to qualify for that claim, state that the label deviates from the requirements of *21 CFR 101.13(e)(2)*, in that it fails to indicate that the food inherently meets the claim and that the claim clearly refers to all foods of that type and not merely to the particular brand to which the labeling is attached.

 Is each ingredient declared in descending order of predominance by weight (21 CFR 101.4(a)(1)), except for ingredients present in amounts of 2 percent or less, which may be placed at the end of the ingredient statement following an appropriate quantifying statement, e.g., "Contains _____ percent or less of _____," or "Less than _____ percent of _____" (the blank percentage to be filled with a threshold level of 2 percent, or if desired, 1.5 percent, 1.0 percent, or 0.5 percent, as appropriate)? (21 CFR 101.4(a)(2))

A. If "*YES*," continue.
B. If "*NO*," state that based on the _____ (insert evidence, e.g., product's formulation, inspection findings, etc.), the ingredients listed as "_____," "_____," and "_____" (insert exact names of the ingredients as declared) are not declared in descending order of predominance by weight.

 Does the product contain ingredients that, themselves, contain two or more ingredients and that have established common or usual names, conform to standards established pursuant to the Meat Inspection or Poultry Products Inspection Acts by the U.S. Department of Agriculture, or conform to definitions and standards of identity established pursuant to the Federal Food, Drug, and Cosmetic Act by one of the two ways, as follows: (21 CFR 101.4(b)(2))

1. *By declaring the established common or usual name of the ingredient, followed by a parenthetical listing of all ingredients contained therein, in descending order of predominance, (21 CFR 101.4(b)(2)(i)) or*
2. *By declaring each ingredient in the multi-component ingredient, in descending order of predominance in the finished food, by its common or usual name, without listing the multi-component ingredient itself? (21 CFR 101.4(b)(2)(ii))*

A. If "*YES*," continue.
B. If "*NO*," for ingredients that are standardized foods, state that the food is fabricated from two or more ingredients, one of which is the standardized multi-component food ingredient _____ (insert the common or usual name of the multi-component ingredient), but the label fails to list each ingredient in the multi-component food ingredient by its common or usual name in accordance with *21 CFR 101.4(b)(2)*.

If "*NO*," for nonstandardized ingredients, state that the food is fabricated from two or more ingredients some of which are the multi-component ingredients "_____," and "_____," etc., (insert the common or usual names of the specific multi-component ingredients as appropriate) but the label fails to list the common or usual name of each ingredient in the multi-component ingredients in accordance with *21 CFR 101.4(b)*.

Q *Does the statement of ingredients include an ingredient percentage statement?*

A. If "YES," the percentage statement must be shown in parentheses followed by the name of the ingredient and expressed in terms of percentage by weight *(21 CFR 101.4(e))*, except where ingredients are grouped together at the end of the ingredient statement, as provided for under *§101.4(a)(2)*. The percentage declaration shall be expressed to the nearest 1 percent *(21 CFR 101.4(e))*.

B. If "*NO*," because the label does not bear a percentage statement in the ingredient statement, continue.

If "*NO*," because percentage is not declared in accordance with *21 CFR 101.4(e)*, state that the ingredient statement bears an ingredient percentage statement, but such statement fails to be declared as required by *21 CFR 101.4(e))*.

Illustration VI.B.3
Information Panel
(21 CFR 101.2)

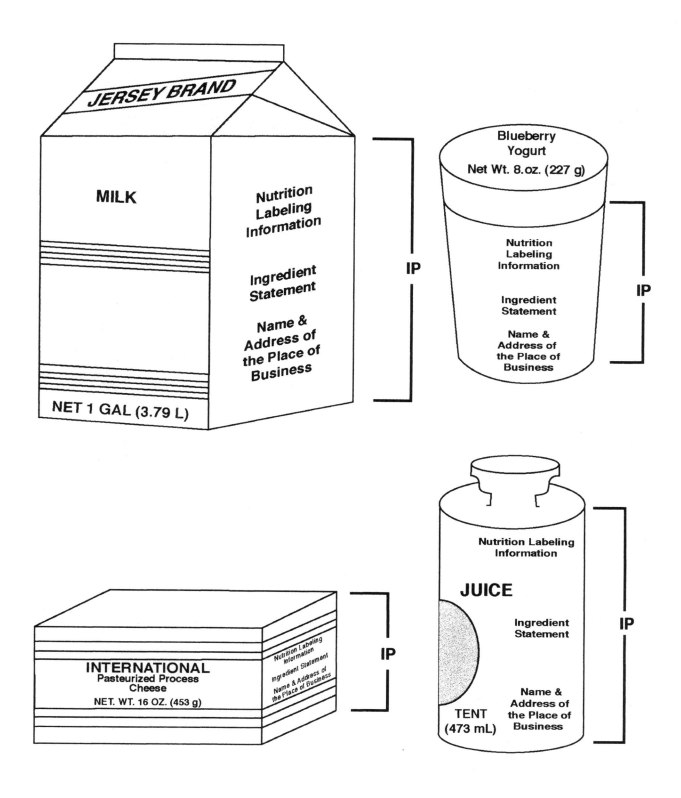

Regulation VI.B.4
Other Common or Usual Names
Specifically Prescribed by Regulations
(21 CFR 101.4(b)(3) through 4(f))

Chapter VI

Food: Designation of Ingredients *(21 CFR 101.4(b)(3) through 4(f))*

b)   *   *   *

(3) Skim milk, concentrated skim milk, reconstituted skim milk, and nonfat dry milk may be declared as "*skim milk*" or "*nonfat milk*".

(4) Milk, concentrated milk, reconstituted milk, and dry whole milk may be declared as "*milk*".

(5) Bacterial cultures may be declared by the word "cultured" followed by the name of the substrate, e.g., "*made from cultured skim milk or cultured buttermilk*".

(6) Sweetcream buttermilk, concentrated sweetcream buttermilk, reconstituted sweetcream buttermilk, and dried sweetcream buttermilk may be declared as "*buttermilk*".

(7) Whey, concentrated whey, reconstituted whey, and dried whey may be declared as "*whey*".

(8) Cream, reconstituted cream, dried cream, and plastic cream (sometimes known as concentrated milk fat) may be declared as "*cream*".

(9) Butteroil and anhydrous butterfat may be declared as "*butterfat*".

(10) Dried whole eggs, frozen whole eggs, and liquid whole eggs may be declared as "*eggs*".

(11) Dried egg whites, frozen egg whites, and liquid egg whites may be declared as "*egg whites*".

(12) Dried egg yolks, frozen egg yolks, and liquid egg yolks may be declared as "*egg yolks*".

(13) [Reserved]

(14) Each individual fat and/or oil ingredient of a food intended for human consumption shall be declared by its specific common or usual name (e.g., "*beef fat*", "*cottonseed oil*") in its order of predominance in the food except that blends of fats and/or oils may be designated in their order of predominance in the foods as "_____ *shortening*" or "*blend of* _____ *oils*", the blank to be filled in with the word "*vegetable*", "*animal*", "*marine*", with or without the terms "*fat*" or "*oils*" or combination of these, whichever is applicable if, immediately following the term, the common or usual name of each individual vegetable, animal, or marine fat or oil is given in parentheses, e.g., "*vegetable oil shortening (soybean and cottonseed oil)*". For products that are blends of fats and/or oils and for foods in which fats and/or oils constitute the predominant ingredient, i.e., in which the combined weight of all fat and/or oil ingredients equals or exceeds the weight of the most predominant ingredient that is not a fat or oil, the listing of the common or usual names of such fats and/or oils in parentheses shall be in descending order of predominance. In all other foods in which a blend of fats and/or oils is used as an ingredient, the listing of the common or usual names in parentheses need not be in descending order of predominance if the manufacturer, because of the use of varying mixtures, is unable to adhere to a constant pattern of fats and/or oils in the product. If the fat or oil is completely hydrogenated, the name shall include the term "*hydrogenated*", or if partially hydrogenated, the name shall include the term "*partially hydrogenated*". If each fat and/or oil in a blend or the blend is completely hydrogenated, the term "*hydrogenated*" may precede the term(s) describing the blend, e.g., "*hydrogenated vegetable oil (soybean, cottonseed, and palm oils)*", rather than preceding the name of each individual fat and/or oil; if the blend of fats and/or oils is partially hydrogenated the term "*partially hydrogenated*" may be used in the same manner. Fat and/or oil ingredients not present in the product may be listed if they may sometimes be used in the product. Such ingredients shall be identified by words indicating that they may not be present, such as "*or*", "*and/or*", "*contains one or more of the following:*" e.g., "*vegetable oil shortening (contains one or more of the following: cottonseed oil, palm oil, soybean oil)*". No fat or oil ingredient shall be listed unless actually present if the fats and/or oils constitute the predominant ingredient of the product as defined in this paragraph (b)(14).

(15) When all the ingredients of a wheat flour are declared in an ingredient statement, the principal ingredient of the flour shall be declared by the name(s) specified in §§*137.105, 137.200, 137.220 and 137.225* of this chapter, i.e., the first ingredient designated in the ingredient list of flour or bromated flour, or enriched flour or self-rising flour is "*flour*", "*white flour*", "*wheat flour*", or "*plain flour*", the first ingredient designated in the ingredient list of durum flour is "*durum flour*"; the first ingredient designated in the ingredient list of whole wheat flour, or bromated

whole wheat flour is "*whole wheat flour*", "*graham flour*", or "*entire wheat flour*"; and the first ingredient designated in the ingredient list of whole durum wheat flour is "*whole durum wheat flour*".

(16) Ingredients that act as leavening agents in food may be declared in the ingredient statement by stating the specific common or usual name of each individual leavening agent in parentheses following the collective name "*leavening*", e.g., "*leavening (baking soda, monocalcium phosphate, and calcium carbonate)*". The listing of the common or usual name of each individual leavening agent in parentheses shall be in descending order of predominance: *Except*, That if the manufacturer is unable to adhere to a constant pattern of leavening agents in the product, the listing of individual leavening agents need not be in descending order of predominance. Leavening agents not present in the product may be listed if they are sometimes used in the product. Such ingredients shall be identified by words indicating that they may not be present, such as "*or*", "*and/or*", or "*contains one or more of the following:*"

(17) Ingredients that act as yeast nutrients in foods may be declared in the ingredient statement by stating the specific common or usual name of each individual yeast nutrient in parentheses following the collective name "*yeast nutrients*", e.g., "*yeast nutrients (calcium sulfate and ammonium phosphate)*". The listing of the common or usual name of each individual yeast nutrient in parentheses shall be in descending order of predominance: *Except*, That if the manufacturer is unable to adhere to a constant pattern of yeast nutrients in the product, the listing of the common or usual names of individual yeast nutrients need not be in descending order of predominance. Yeast nutrients not present in the product may be listed if they are sometimes used in the product. Such ingredients shall be identified by words indicating that they may not be present, such as "*or*", "*and/or*", or "*contains one or more of the following:*"

(18) Ingredients that act as dough conditioners may be declared in the ingredient statement by stating the specific common or usual name of each individual dough conditioner in parentheses following the collective name "*dough conditioner*", e.g., "*dough conditioners (L-cysteine, ammonium sulfate)*". The listing of the common or usual name of each dough conditioner in parentheses shall be in descending order of predominance: *Except*, That if the manufacturer is unable to adhere to a constant pattern of dough conditioners in the product, the listing of the common or usual names of individual dough conditioners need not be in descending order of predominance. Dough conditioners not present in the product may be listed if they are sometimes used in the product. Such ingredients shall be identified by words indicating that they may not be present, such as "*or*", "*and/or*", or "*contains one or more of the following:*"

(19) Ingredients that act as firming agents in food (e.g., salts of calcium and other safe and suitable salts in canned vegetables) may be declared in the ingredient statement, in order of predominance appropriate for the total of all firming agents in the food, by stating the specific common or usual name of each individual firming agent in descending order of predominance in parentheses following the collective name "*firming agents*". If the manufacturer is unable to adhere to a constant pattern of firming agents in the food, the listing of the individual firming agents need not be in descending order of predominance. Firming agents not present in the product may be listed if they are sometimes used in the product. Such ingredients shall be identified by words indicating that they may not be present, such as "*or*", "*and/or*", "*contains one or more of the following:*"

(20) For purposes of ingredient labeling, the term "*sugar*" shall refer to sucrose, which is obtained from sugar cane or sugar beets in accordance with the provisions of *§184.1854* of this chapter.

(21) [Reserved]

(22) Wax and resin ingredients on fresh produce when such produce is held for retail sale, or when held for other than retail sale by packers or repackers shall be declared collectively by the phrase "*coated with food-grade animal-based wax, to maintain freshness*" or the phrase "*coated with food grade vegetable-, petroleum-, beeswax-, and/or shellac-based wax or resin, to maintain freshness*" as appropriate. The terms "*food-grade*" and "*to maintain freshness*" are optional. The term "*lac-resin*" may be substituted for the term "*shellac.*"

(23) When processed seafood products contain fish protein ingredients consisting of the

myofibrillar protein fraction from one or more fish species and the manufacturer is unable to adhere to a constant pattern of fish species in the fish protein ingredient, because of seasonal or other limitations of species availability, the common or usual name of each individual fish species need not be listed in descending order of predominance. Fish species not present in the fish protein ingredient may be listed if they are sometimes used in the product. Such ingredients must be identified by words indicating that they may not be present, such as "*or*", "*and/or*", or "*contains one or more or the following:*" Fish protein ingredients may be declared in the ingredient statement by stating the specific common or usual name of each fish species that may be present in parentheses following the collective name "*fish protein*", e.g., "*fish protein (contains one of more of the following: Pollock, cod, and/or Pacific whiting)*".

c) When water is added to reconstitute, completely or partially, an ingredient permitted by paragraph (b) of this section to be declared by a class name, the position of the ingredient class name in the ingredient statement shall be determined by the weight of the unreconstituted ingredient plus the weight of the quantity of water added to reconstitute that ingredient, up to the amount of water needed to reconstitute the ingredient to single strength. Any water added in excess of the amount of water needed to reconstitute the ingredient to single strength shall be declared as "*water*" in the ingredient statement.

d) When foods characterized on the label as "*nondairy*" contain a caseinate ingredient, the caseinate ingredient shall be followed by a parenthetical statement identifying its source. For example, if the manufacturer uses the term "*nondairy*" on a creamer that contains sodium caseinate, it shall include a parenthetical term such as "*a milk derivative*" after the listing of sodium caseinate in the ingredient list.

e) If the percentage of an ingredient is included in the statement of ingredients, it shall be shown in parentheses following the name of the ingredient and expressed in terms of percent by weight. Percentage declarations shall be expressed to the nearest 1 percent, except that where ingredients are present at levels of 2 percent or less, they may be grouped together and expressed in accordance with the quantifying guidance set forth in paragraph (a)(2) of this section.

f) Except as provided in *§101.100*, ingredients that must be declared on labeling because there is no label for the food, including foods that comply with standards of identity, shall be listed prominently and conspicuously by common or usual name in the manner prescribed by paragraph (b) of this section.

Chapter VI

Section C: Name and Place of Business *(21 CFR 101.5)*

1. Summary of the Requirements

a) The label of a food in packaged form shall specify conspicuously the name and place of business of the manufacturer, packer, or distributor. The requirement with respect to the "*name*" of the manufacturer, packer, or distributor will be deemed to be satisfied if:
   (i) In the case of a corporation, only the actual corporate name is used, which may be preceded or followed by the name of the particular division of the corporation; and
   (ii) In the case of an individual, partnership, or association, the name under which the business is conducted is used.

b) Where the food is not manufactured by the person whose name appears on the label, the name shall be qualified by a phrase that reveals the connection such person has with such food; such as "*Manufactured for _____,*" "*Distributed by _____,*" or any other wording that expresses the facts.

c) The statement shall include the street address, city, state, and ZIP code; however, the street address may be omitted if it is shown in a current city directory or telephone directory. In the case of nonconsumer packages, the ZIP code shall appear either on the label or the labeling (including invoice).

d) If a person manufactures, packs, or distributes a food at a place other than his principal place of business, he may declare that principal place of business on the label in lieu of the actual place where such food was manufactured, or packed, or is to be distributed, unless such statement would be misleading.

2. Questions and Responses

The questions and responses sections included under Chapter VI are a series of established requirements designed to aid the reviewers in establishing the degree of compliance that a specific food label and its labeling comply with applicable laws and regulations. It is not intended to represent all questions that may arise during the review process, but as examples of typical questions during a label review. This section can also serve as a teaching aid for less experienced reviewers and as a ready reference for experienced reviewers. It can also serve as an example for developing a response to advise those responsible for food labels and labeling of needed corrections.

Does the label bear the name and business address of the product's manufacturer, packer, or distributor?

A. If "*YES,*" continue.
B. If "*NO,*" state that the label fails to bear the name and address of the food's manufacturer, packer, or distributor, as required by *21 CFR 101.5(a).*

Is the statement containing the name and place of business conspicuously declared on the label?

[NOTE: In the case of a corporation, the name may be preceded or followed by the name of the particular division of the corporation; or in the case of an individual, partnership, or association, the name under which the business is conducted shall be used. (21 CFR 101.5(b))]

A. If "*YES,*" continue.
B. If "*NO,*" state that the responsible firm's name and address fail to appear conspicuously on the label, as required by *21 CFR 101.5(b).*

If the food is manufactured by a firm whose name does not appear on the label, is the name of the firm on the label qualified by a phrase that reveals such firm's connection with the product, e.g., "Manufactured for _____," "Distributed by _____," or any other wording expressing the fact?

A. If "*YES*," continue.
B. If "*NO*," state that the product is not manufactured by the firm whose name appears on the label, and that the label fails to reveal the connection of the listed firm and the product, as required by *21 CFR 101.5(c)*.

 Does the place of business statement include the street address, city, state, and ZIP code?

A. If "*YES*," continue.
B. If "*NO*," but is required, state that the label fails to bear _____ (insert the missing component(s), e.g., street address, city, state, and/or ZIP code), as required by *21 CFR 101.5(d)*.

Chapter VI

Section D: Net Quantity of Contents Statements *(21 CFR 101.105)*

1. Summary of the Requirements

a) The net quantity of contents statement shall be on the principal display panel and alternate principal display panels of the label.

b) The statement shall be located in the lower 30% of the principal display panel. Packages with 5 square inches or less are exempted from this requirement.

c) The statement shall be declared in lines generally parallel to the base on which the package rests.

d) Statement shall be a distinct item on the principal display panel (separated from information appearing above and below the statement by a space equal to the height of the lettering used in the statement, and separated from information to the left or right of the statement by a space equal to twice the width of the letter "N").

e) The statement shall be expressed in terms of:
   (i) Weight, if the food is solid, semisolid, or viscous, or a mixture of solid and liquid;
   (ii) Fluid measure, if liquid, except if there is a general consumer usage and trade custom of declaring the contents of a liquid by weight, or a solid, semisolid, or viscous product by fluid measure, such measure may be used; or
   (iii) Numeral count, except that when count does not give adequate information, a combination of numerical count and weight, measure, or size of the individual units shall be used.

f) The statement of weight shall be in terms of avoirdupois pound and ounce.

g) The statement of fluid measure shall be in terms of U.S. gallon and quart, pint, and fluid ounce subdivisions thereof.

h) In the case of frozen food that is sold and consumed in a frozen state, the volume shall be expressed at the frozen temperature.

i) In the case of refrigerated food that is sold in the refrigerated state, the volume shall be expressed at 40 F (4 C).

j) In the case of other foods, the volume shall be expressed at 68 F (20 C).

k) Statements of dry measure shall be in terms of the U.S. bushel of 2,150.42 cubic inches and peck, dry quart, and dry pint subdivisions thereof.

l) Statement with common fractions shall be expressed in term halves, quarters, eighths, sixteenths, or thirty-seconds, except if there exists a firmly established consumer usage and trade custom employing different common fractions in the net quantity declaration.
   (i) A common fraction shall be reduced to its lowest term, e.g., Net Wt. 24 oz (1½ lb).
   (ii) Statement with decimal fractions shall not be carried out to more than three places, e.g., Net Wt 24 oz (1.50 lb).
   (iii) A statement that includes small fractions of an ounce shall be deemed to permit smaller variations than a statement that does not include such fractions.

m) Both Metric and U.S. Customary System declarations are required, e.g., Net Wt. 1 lb 8 oz (680 g), 500 ml (1 pt 0.9 fl oz).

n) The statement shall accurately reveal the quantity of food in the package, exclusive of the wrapper and packaging material.

o) The statement for products designed to deliver under pressure shall state the net quantity of contents that will be expelled when the instructions for use on the container are followed. The propellant is included in the net quantity of contents declaration.

p) The statement must appear prominently and conspicuously, in easily legible boldface type in distinct contrast to other matter on the package, except when blown, embossed, or molded on a glass or plastic surface.

q) The requirements for conspicuousness and legibility shall include the specification that:
   (i) The ratio of the height to the width of a letter should not exceed a differential of 3 units to 1 unit (i.e., not more than 3 times high as it is wide);
   (ii) Letter heights pertain to upper case or capital letters. When upper and lower case letters are used, the lower case letter "o", or its equivalent, is the minimum standard; and
   (iii) When fractions are used, each component numeral shall meet one-half minimum height standards.

r) The statement shall be established in relation to the area of the principal display panel.

Minimum Type Size	Area of Principal Display Panel
1/16 inch in height	5 square inches or less
1/8 inch in height	more than 5 square inches but not more than 25 square inches
3/16 inch in height	more than 25 square inches but not more than 100 square inches
1/4 inch in height	more than 100 square inches but not more than 400 square inches
1/2 inch in height	more than 400 square inches

s) On a multiunit retail package (two or more individually packaged units of the identical commodity in the same quantity, intended to be sold as part of multiunit package but capable of being sold individually), the net quantity of contents statement shall appear on the outside of the package and include the following components:

(i) The number of individual units in the package;

(ii) The quantity of each individual unit; and in parentheses; and

(iii) The total quantity of contents of multiunit package.

Examples of net quantity of contents statements for multiunit packages:
6-16 oz bottles – (96 fl oz);
3-16 oz cans – (net wt 48 oz).

2. Questions and Responses

The questions and responses sections included under Chapter VI are a series of established requirements designed to aid the reviewers in establishing the degree of compliance that a specific food label and its labeling comply with applicable laws and regulations. It is not intended to represent all questions that may arise during the review process, but as examples of typical questions during a label review. This section can also serve as a teaching aid for less experienced reviewers and as a ready reference for experienced reviewers. It can also serve as an example for developing a response to advise those responsible for food labels and labeling of needed corrections.

[NOTE: Nothing in this section shall prohibit supplemental statements at locations other than the principal display panel(s). Such supplemental statements shall not include any term qualifying a unit of weight, mass, measure, or count that tends to exaggerate the amount of the food in the package, e.g., "jumbo quart" and "full gallon." Dual or combination declarations of net quantity of contents such as net weight plus numerical count, net contents plus dilution directions of a concentrate, etc., are not regarded as supplemental net quantity of contents statements and may be located on the principal display panel. (21 CFR 101.105(o))

For quantities, the following abbreviations may be used (period and plural forms are optional): weight - wt, ounces - oz, pound - lb, gallon - gal, pint - pt, quart - qt, fluid - fl. (21 CFR 101.105(n))

The statement shall be accurate; however, reasonable variations caused by loss or gain of moisture during the course of good distribution practice or unavoidable deviations in good manufacturing practice will be recognized. (21 CFR 101.105(q))

Net quantity of contents declaration on pickles and pickle products, including relishes but excluding two whole pickles in clear plastic bags which may be declared by count, must be expressed in terms of the U.S. gallon of 231 cubic inches and quart, pint, and fluid ounce subdivisions. (21 CFR 101.105(r))]

Is the product in package form?

A. If "*YES*," continue.

B. If "*NO*," STOP HERE. A net quantity of contents statement is not required.

Chapter VI

Does the label bear a net quantity of contents statement?

A. If "*YES*," continue.
B. If "*NO*," state that the product is in package form, but the label fails to bear a net quantity of contents statement as required by *21 CFR 101.105(a)*.

Is the net quantity of contents statement declared in both customary inch-pound and SI (metric) units?

A. If "*YES*," continue.
B. If "*NO*," state that the net quantity of contents statement(s) fails to be declared in both customary inch pound and metric units.

Does the net quantity of contents statement appear on the principal display panel and each alternate principal display panel?

A. If "*YES*," continue.
B. If "*NO*," state that the net quantity of contents statement fails to appear on the _____ panel(s) of the label (the blank to be filled with the appropriate panel(s) where the net quantity of contents statement is missing, e.g., alternate principal display panel(s)). *(21 CFR 101.105(a))*

Does the net quantity of contents statement appear as a distinct item on the principal display and alternate principal display panels, separated from other printed matter on the panel?

A. If "*YES*," continue.
B. If "*NO*," state that the net quantity of contents statement fails to appear as a distinct item on the principal display panel(s) due to the failure of the statement to be separated by at least a space equal to _____ (the blank is to be filled with the statement "twice the width of the letter 'N' of the style of type used in the quantity of contents statement from other printed label information appearing to the left or right of the declaration" or "the height of the lettering used in the declaration from other printed label information appearing above or below the declaration," as applicable). *(21 CFR 101.105(f))*

Does the statement exclude terms qualifying a unit of weight, measure, or count (such as "jumbo quart" and "full gallon") that tends to exaggerate the amount of the food in the container? *(21 CFR 101.105(o))*

A. If "*YES*," continue.
B. If "*NO*," state that the net quantity of contents statement bears the term "_____" (insert the term used in the declaration that tends to exaggerate the amount of food in the container) that tends to exaggerate the amount of food in the container.

Is the statement presented in the lower 30 percent of the area of the principal display panel in lines generally parallel to the base on which the package rests as it is designed to be displayed?

*[**NOTE:** Packages having a principal display panel of 5 square inches or less are not subject to the 30 percent requirement. (21 CFR 101.105(f))]*

A. If "*YES*," continue.
B. If "*NO*," state that the net quantity of contents statement fails to appear in the lower 30 percent of the principal display panel and/or in lines generally parallel to the base of the package on which it rests, as applicable.

Does the statement accurately reveal the quantity of food (net quantity) in the package, exclusive of wrappers and other material packed with the food?

[NOTE: In case of foods packed in containers designed to deliver the food under pressure, the declaration shall state the net quantity of contents that will be expelled when the instructions for use as shown on the container are followed.

The propellant is included in the net quantity of contents declaration *(21 CFR 101.105(g))]*

A. If "*YES*," continue.
B. If "*NO*," where the sample shows an average (48 units examined) short weight of one percent or more, the product is misbranded within the meaning of section 403(e)(2) of the Act in that the label fails to bear an accurate statement of the net quantity of contents of the container.

Does the declaration appear in conspicuous and easily legible boldface print or type in distinct contrast (by typography, layout, color embossing, or molding) to other matter on the package?

[NOTE: An exception to this requirement is made when the declaration of net quantity is blown, embossed, or molded on a glass or plastic surface when all label information is so formed on the surface. (21 CFR 101.105(h))]

Requirements of conspicuousness and legibility shall include the following specifications:
*   The ratio of the height to the width of the letters shall not exceed a differential of 3 units to 1 unit (i.e., no more than 3 times as high as it is wide);
*   Letter heights pertain to upper case or capital letters;
*   When upper and lower case or all lower case letters are used, it is the lower case letter "o" or its equivalent that shall meet the minimum standards; and
*   When fractions are used, each component numeral shall meet one-half the minimum height standards. *(21 CFR 101.105(h))*

A. If "*YES*," continue.
B. If "*NO*," state that the net quantity of contents statement fails to be declared conspicuously on the label due to _____ (insert the specific reasons the statement is inconspicuous, e.g., the lack of contrast between the type used in the declaration and the background color).

Are the numbers and letters in the declaration in a type size established in relationship to the area of the principal display panel of the package and uniform for all packages of substantially the same size by complying with the following type size specifications?

*   If the area of the principal display panel is 5 square inches or less, the net quantity of contents statement shall be not less than one-sixteenth inch in height.
*   If the area is more than 5 square inches but not more than 25 square inches, the declaration shall not be less than one-eighth inch in height.
*   If the area is more than 25 square inches but not more than 100 square inches, the declaration shall be not less than three-sixteenths inch in height.
*   If the area is more than 100 square inches but not more than 400 square inches, the declaration shall be not less than one-fourth inch in height.
*   If the area is more than 400 square inches, the declaration shall be not less than one-half inch in height. *(21 CFR 101.105(i))*

A. If "*YES*," continue.
B. If "*NO*," state that the type size of the declaration fails to be established in relationship to the area of the principal display panel as required by *21 CFR 101.105(i).*

Chapter VI

Is the statement declared in terms of weight for a product that is solid, semisolid, or viscous or a mixture of solid and liquid; in terms of fluid measure if the food is a liquid; or in terms of U.S. bushel of 2,150.42 cubic inches and peck, dry quart, and dry pint subdivisions thereof for dry measure, unless there is a firmly established general consumer usage and trade custom to the contrary? *(21 CFR 101.105(a))*

A. If "*YES,*" continue.

B. If "*NO,*" state that the product is a _____, but the net quantity of contents statement fails to be declared in terms of _____ (the first blank to be filled with state of the product, e.g., solid; and the second blank to be filled with required measure, e.g., weight).

If the net quantity of contents is in terms of numerical count, does the declaration give adequate information as to the quantity of the food in the package?

A. If "*YES,*" continue.

B. If "*NO,*" state that the numerical count declaration fails to give adequate information relative to the net quantity of contents of the food in the package. The numerical count declaration should be combined with a statement of _____ so as to provide adequate information concerning the quantity of food in the package (the blank to be filled with a statement of weight, measure, or size of the individual units of the foods, as appropriate). *(21 CFR 101.105(c))*

Does the statement contain common or decimal fractions such as halves, quarters, eighths, sixteenths, or thirty-seconds in compliance with 21 CFR 101.105(d)?

[NOTE: If there exists a firmly established general consumer usage and trade custom of employing different common fractions in the net quantity of contents declaration for a particular commodity, it may used. (21 CFR 101.105(d))]

A. If "YES," continue.

B. If "NO," because the statement does not contain fractions, continue.

If "NO," because the fractions used are not common fractions such as halves, quarters, eighths, sixteenths, or thirty-seconds, or if different fractions are used that were not established through general consumer usage and trade custom, state that the fraction in the net quantity of contents declaration fails to be declared in terms of common or decimal fractions in accordance with *21 CFR 101.105(d).*

Is the product in a multiunit retail package (a package containing two or more individually packaged units of the identical commodity and in the same quantity, intended to be sold as part of the multiunit retail package, but capable of being individually sold in full compliance with all requirements of the regulations) that bears a net quantity of contents statement on the outside of the package? *(21 CFR 101.105(s))*

A. If "YES," continue.

[NOTE: The statement shall include the number of individual units, the quantity of each individual unit, and, in parentheses, the total quantity of contents of the multiunit package.

The declaration of total quantity may be preceded by the term "total" or the phrase "total contents."

The declaration shall appear in both customary inch-pound and SI (metric) units.]

B. If "NO," because the product is not in a multiunit package, continue.

If "NO," because it is a multiunit retail package and is not in compliance with the above requirements, state the deviation and the specific regulation with which the label fails to comply.

Chapter VI

Section E: Nutrition Labeling *(21 CFR 101.9)*

1. **Summary of the Requirements**

   a) Food products in package form that are offered for sale for human consumption shall bear nutrition labeling, except when exempted by *§101.9(j)*. (See Regulations VI.E.15)

   b) Food products in package form that are required to bear nutrition labeling information shall bear such information on the label in a format specified in *§101.9(d)*. (See Illustrations VI.E.4(a) through 4(m))

   c) In the case of food products that are not in package form and that are intended for human consumption, the required nutrition information shall be clearly displayed at the point of purchase by use of counter cards, signs, or tags affixed to the products. Alternatively, some other appropriate device such as a booklet or loose-leaf binder containing the information or other appropriate format that is available at the point of purchase may be used.

   d) Solicitation of requests for nutrition information by a statement "For nutrition information write to _____" on the label or in the labeling of a food, or providing such information in a direct written reply to a solicited or unsolicited request, does not subject the label or the labeling of the food that is exempted from nutrition labeling under *§101.9(j)* to the nutrition labeling requirements, if the reply to the request conforms to *§101.9*.

   e) If any vitamin or mineral is added to a food so that a single serving provides 50 percent or more of the Reference Daily Intake (RDI) for the age group for which the product is intended, unless such addition is permitted by or required in other regulations, e.g., a standard of identity or nutritional quality guideline, or is otherwise exempted, the food shall be considered a food for special dietary use within the meaning of *21 CFR 105.3(a)(1)(iii)*. (See Regulations VII.A)

   f) All nutrient and food component quantities that are required to be declared shall be declared in relation to a serving, except as provided in *§101.9(h)(3)*. (See Regulations VI.E.7)

   g) The term "*Serving*" or "*Serving Size*" means an amount customarily consumed per eating occasion by persons 4 years of age or older which is expressed in a common household measure that is appropriate to the food. When a food is specially formulated for infants or processed for use by infants or by toddlers, a serving or serving size means an amount of food customarily consumed per eating occasion by infants up to 12 months of age or by children 1 through 3 years of age, respectively.

      (i) The serving size declared on the label shall be determined from the "Reference Amount Customarily Consumed Per Eating Occasion" (reference amounts) as required by *21 CFR 101.12(b)*. (See Chart VI.E.5 & E.6)

      (ii) When food products are intended for weight control and are available only through a weight-control program, the manufacturer may determine the serving size that is consistent with the meal plan of the program. Such products must bear a statement, "for sale only through the _____ program" (fill in the blank with the name of the appropriate weight-control program, e.g., Smith's Weight Control Program), on the principal display panel. However, the reference amounts in *§101.12(b)* shall be used for purposes of evaluating whether weight-control products that are available only through a weight-control program qualify for nutrient content claims or health claims. *(§101.9(b)(2))*

      (iii) When the serving size, determined from the reference amount in *§101.12(b)* and the procedures described in the serving size regulations, falls exactly half way between two serving sizes, e.g., 2.5 tbsp., manufacturers shall round the serving size up to the next incremental size. *(§101.9(b)(5)(ix))*

   h) For nutrition labeling purposes, a teaspoon (tsp) means 5 milliliters (mL), a tablespoon (tbsp) means 15 mL, a cup means 240 mL, 1 fluid ounce (fl oz) means 30 mL, and 1 ounce (oz) in weight means 28 grams (g). *(§101.9(b)(5)(viii))*

   i) A label statement regarding a serving shall be the serving size expressed in common household measures and shall be followed by the equivalent metric quantity in parenthesis (fluids in milliliters and all other foods in grams) except for single-serving containers.

      (i) For single-serving containers, the parenthetical metric quantity, which will be presented as part of the net weight statement on the principal display panel, is not required on a drained weight basis according to *§101.9(b)(9)*.

    (ii) If a manufacturer voluntarily provides the metric quantity on products that can be sold as single servings, then the numerical value provided as part of the serving size declaration must be identical to the metric quantity declaration provided as part of the net quantity of contents statement. *(§101.9(b)(7)(i))*

(j) The declaration of nutrient and food component content shall be on the basis of food as packaged or purchased with the following exceptions: (1) raw fish covered under *21 CFR 101.42*, (2) packaged single-ingredient products that consist of fish or game meat as provided for in *§101.9(j)(11)*, and (3) foods that are packed or canned in water, brine, or oil but whose liquid packing medium is not customarily consumed (e.g., canned fish, maraschino cherries, pickled fruits, and pickled vegetables). Declaration of the nutrient and food component content of raw fish shall follow the provisions in *21 CFR 101.45*. Declaration of the nutrient and food component content of foods that are packed in liquid which is not customarily consumed shall be based on the drained solids. *(§101.9(b)(9))*

(k) Another column of figures may be used to declare the nutrient and food component information:
    (i)   Per 100 g or 100 mL, or per 1 oz or 1 fl oz of the food as packaged or purchased;
    (ii) Per 1 unit if the serving size of a product in discrete units in a multiserving container is more than 1 unit; and
    (iii) Per cup popped for popcorn in a multiserving container. *(§101.9(b)(10))*

(l) If a product is promoted on the label, labeling, or advertising for a use that differs in quantity by twofold or greater from the use upon which the reference amount in *§101.12(b)* was based (e.g., liquid cream substitutes promoted for use with breakfast cereals), the manufacturer shall provide a second column of nutrition information based on the amount customarily consumed in the promoted use, in addition to the nutrition information per serving derived from the referenced amount in *§101.12(b),* except that nondiscrete bulk products that are used primarily as ingredients (e.g., flour, sweeteners, shortenings, oils), or traditionally used for multipurposes (e.g., eggs, butter, margarine), and multipurpose baking mixes are exempt from this requirement. *(§101.9(b)(11))*

(m) The declaration of nutrition information on the label and in labeling of a food shall contain information about the level of those nutrients listed in Chart VI.E.12, except for those nutrients whose inclusion, and the declaration of amounts, is voluntary as set forth in the Chart. No nutrients or food components other than those listed in the above-mentioned Chart as either mandatory or voluntary may be included within the nutrition label. The declaration of nutrition information may be presented in the simplified format set forth in *§101.9(f)*. (See Illustration VI.E.4(x))

(n) Nutrient information shall be presented using the nutrient names specified in *§101.9(c)* (See Chart VI.E.12) and in the order and formats specified in *§101.9(d)* or *(e)*. (See Illustrations VI.E.4(a) and 4(m)) In the interest of uniformity of presentation FDA urges that the nutrition information be presented using the graphic specifications set forth in Appendix B to part 101. (See Chart VI.E.13 & 14)
    (i)   The nutrition information shall be set off in a box by use of hairlines and shall be all black or one color type, printed on a white or other neutral contrasting background whenever practical. (See Illustration VI.E.4(a))
    (ii) The information shall be presented under the identifying heading of "Nutrition Facts" which shall be set in a type size larger than all other print size in the nutrition label and, except for labels presented according to the format provided for in *§101.9(d)(11),* unless impractical, shall be set the full width of the information provided under *§101.9(d)(7)*.

(o) Nutrition information may be presented for two or more forms of the same food (e.g., both "as purchased" and "as prepared") or for common combinations of food as provided for in *§101.9(h)(4)* (See Illustration VI.E.4(c)), for different units (e.g., slices of bread or per 100 grams) as provided for in *§101.9(b)*, or two or more groups for which RDI's are established (e.g., both infants and children less than 4 years of age). When such dual labeling is provided, equal prominence shall be given to both sets of values. The information shall be presented in a format consistent with the requirements of *§101.9(d)*, except that;
    (i)   Following the subheading of "Amount Per Serving," there shall be two or more column headings accurately describing the forms of the same food (e.g., "Mix" and "Baked"), the

combinations of food, the units, or the RDI groups that are being declared. The column representing the product as packaged, and according to the label serving size based on the reference amount in *§101.12(b)*, shall be to the left of the numeric columns.

(ii) When the dual labeling is presented for two or more forms of the same food, for combinations of food, or for different units, total calories and calories from fat (and calories from saturated fat, when declared) shall be listed in a column and indented as specified in *§101.9(d)(5)* with quantitative amounts declared in columns aligned under the column headings as set forth in *§101.9(e)(1)*.

(iii) Quantitative information by weight required in *§101.9(d)(7)* shall be specified for the form of the product as packaged and according to the labeled serving size based on the reference amount in *§101.12(b)*.

    (a) Quantitative information by weight may be included for other forms of the product represented by the additional column(s) either immediately adjacent to the required quantitative information by weight for the product as packaged and according to the labeled serving size based on the reference amount in *§101.12(b)* or as a footnote.

    (b) Total fat and its quantitative amount by weight shall be followed by an asterisk (or other symbol) (e.g., "Total fat (2 g)*") referring to another asterisk (or symbol) at the bottom of the nutrition label identifying the form(s) of the product for which the nutrition information is presented.

(iv) Information required under *§101.9(d)(7)(ii)* and *(d)(8)* shall be presented under the subheading "% DAILY VALUE" and in columns directly under the column headings as set forth in *§101.9(e)(1)*.

2. Questions and Responses

The questions and responses sections included under Chapter VI are a series of established requirements designed to aid the reviewers in establishing the degree of compliance that a specific food label and its labeling comply with applicable laws and regulations. It is not intended to represent all questions that may arise during the review process, but as examples of typical questions during a label review. This section can also serve as a teaching aid for less experienced reviewers and as a ready reference for experienced reviewers. It can also serve as an example for developing a response to advise those responsible for food labels and labeling of needed corrections.

Is the product exempted from nutrition labeling? *(See Regulations VI.E.15)*

[NOTE: The solicitation of requests for nutrition information by use of the statement "For nutrition information write to _____" on the label or in labeling of a food, or providing such information in a direct reply to a solicited or unsolicited request, does not subject the label or labeling of a food that is exempted from nutrition labeling under 21 CFR 101.9(j) to the provisions of §101.9, if the reply to the request conforms to these requirements. (21 CFR 101.9(a)(3))]

    A. If "*YES*," STOP HERE, nutrition labeling is not required.
    B. If "*NO*," or if exempted, and the label bears a nutrient content claim, or if the label bears nutrition information, continue.

Is the food offered for sale in package form?

    A. If "*YES*," nutrition information is required to appear on the label in a format specified by regulation, continue.
    B. If "*NO*," nutrition information shall be displayed clearly at the point of purchase, e.g., on a counter card, sign, or tag affixed to the product, or some other appropriate format or device such as in a booklet, or loose-leaf binder, etc. *(21 CFR 101.9(a)(2)).*

Does the package have more than 40 square inches with at least three continuous vertical inches of space available for labeling?

A. If "*YES,*" vertical formats may be used where appropriate:
- Full vertical format. *(21 CFR 101.9(d)(12))* (See Illustration VI.E.4(a))
- Full vertical format with footnote to the side. (See Illustration VI.E.4(b))
- Full vertical - dual declaration, "As Packaged" and "As Prepared". *(21 CFR 101.9(e))* (See Illustration VI.E.4(c))
- Full vertical - dual declaration, combination with another food. *(21 CFR 101.9(e)(2))* (See Illustration VI.E.4(c))
- Full vertical bilingual declaration. *(21 CFR 101.9(d)(14))* (See Illustration VI.E.4(d))
- Full vertical aggregate declaration. *(21 CFR 101.9(d)(12)(ii))* (See Illustration VI.E.4(e))
- Simplified vertical format. *(21 CFR 101.9(f))* (See Illustration VI.E.4(f))
- Shortened vertical format. *(FR., Vol. 58, 1/6/93, 2084 and 2192)* (See Illustration VI.E.4(g))

B. If "*NO,*" state that the label fails to bear nutrition labeling in the applicable format prescribed by *21 CFR 101.9.*

Does the package have 40 square inches or more total surface area but does not have at least three continuous vertical inches of space available for labeling, or does the package have less than 40 square inches but more than 12 square inches of space available for labeling?

A. If "*YES,*" the following formats may be used where appropriate:
- Simplified vertical format. *(21 CFR 101.9(f))* (See Illustration VI.E.4(f))
- Shortened vertical format. (See Illustration VI.E.4(g))
- Full tabular format. (See Illustration VI.E.4(h))
- Tabular format for small packages. (See Illustration VI.E.4(i))
- Simplified tabular format. (See Illustration VI.E.4(j))

B. If "*NO,*" the linear format (full linear, shortened linear or simplified linear) may be used on packages with 40 or less square inches of available space, only if the label cannot accommodate the tabular format. *(21 CFR 101.9(j)(13)(ii)(A))* (See Illustration VI.E.4(k))

Does the package have less than 12 square inches of space available for labeling?

A. If "*YES,*" the label is exempted from nutrition labeling unless the label bears nutrition claims or other nutrition information in any context. *(21 CFR 101.9(j)(13)(i))*

[**NOTE**: *If the above exemption is used, the label shall bear an address or telephone number that a consumer can use to obtain the required nutrition information (e.g., for nutrition information, call 1-800-123-4567). (21 CFR 101.9(j)(13)(i)(A))*

When nutrition labeling is provided, voluntary or otherwise, all required nutrition information shall be in type size no smaller than 6 point or all upper case type of 1/16 inch minimum height, except that individual serving size packages of food served with meals in restaurants, and not intended for sale at retail, may comply with 21 CFR 101.2(c)(5) and 21 CFR 101.9(j)(13)(i)(B).]

B. If "*NO,*" because the package has more than 12 square inches of space available for labeling, it should bear nutrition labeling as previously discussed, unless otherwise exempted.

Is the food represented as food for children?

A. If "*YES,*" then the format for nutrition labeling shall be in compliance with the following regulations:
- Children less than 2 years of age. *(21 CFR 101.9(j)(5)(i))* (See Illustration VI.E.4(l))
- Children less than 4 years of age. *(21 CFR 101.9(j)(5)(ii))* (See Illustration VI.E.4(m))

If "YES," but the format does not comply with the requirements above, state that the food is represented for children less than _____ (insert age group represented) years of age, but fails to bear nutrition labeling in the format prescribed by _____ (insert CFR reference).

B. If "*NO*," it is not represented as a food for children less than 4 years of age, continue.

Q *Is the nutrition information set off in a box with hairlines and printed in all black letters and numbers, or one color type, on a white or other neutral contrasting background? (21 CFR 101.9(d)(1)(i))*

A. If "*YES*," continue.

B. If "NO," state that the nutrition information fails to _____ (insert the specific deviations from the requirements) in accordance with *21 CFR 101.9(d)(1)(i)*.

Q *Are nutrients and components declared in relation to the specific format used? (21 CFR 101.9(d)(1))*

A. If "*YES*," continue.

B. If "*NO*," state that the nutrition information fails to be declared in the required format.

Q *Are quantities of all nutrients and food components declared in relation to an appropriate serving of the food? (21 CFR 101.9(b) and 101.12). (See Illustrations VI.E.5, VI.E.6, VI.E.8, and VI.E.10))*

A. If "*YES*," continue.

B. If "*NO*," state that the quantity of the component _____ (insert the specific nutrient/food component that deviates from the requirement) in the Nutrition Facts box fails to be declared in relation to an appropriate serving of the food.

Q *Does the nutrition information in the Nutrition Facts box use a single easy-to-read type style? (21 CFR 101.9(d)(1)(ii)(A))*

A. If "*YES*," continue.

B. If "*NO*," state that the _____ (insert the information that deviates from the requirement) in the Nutrition Facts box is not declared in a single easy-to-read type style.

Q *Are upper and lower case letters used in the box? (21 CFR 101.9(d)(1)(ii)(B))*

A. If "*YES*," continue.

B. If "*NO*," state that the declaration for _____ (insert the specific declaration(s) not appropriately declared) in the Nutrition Facts box fails to be declared in upper and lower case letters.

Q *Does the space between two lines of text for all information in the Nutrition Facts box have the required point leading? (21 CFR 101.9(d)(1)(ii)(C)) (See Illustration VI.E.13) and Chart VI.E.14*

A. If "*YES,*" continue.

B. If "*NO*," state that the letters used in the words _____ and _____ touch and therefore, deviate from the requirements of *21 CFR 101.9(d)(1)(ii)(D)*.

Q *Is the type size for all information in the Nutrition Facts box in keeping with the requirements under 21 CFR 101.9(d)(1)(iii) and (iv)?*

A. If "*YES*," continue.

B. If "*NO*," state that the type size for the _____ declaration fails to be declared in the size type required by *21 CFR 101.9(d)(iii)* or *(iv)*, as appropriate.

Q *Are the headings "Nutrition Facts," "Amount Per Serving," "% Daily Value," "Calories," "Total Fat," "Cholesterol," "Sodium," "Total Carbohydrate," "Protein," and the percentage amount required by paragraph 21 CFR 101.9(d)(7)(ii) highlighted by bold or extra bold type or other highlighting? (21 CFR 101.9(d)(1)(iv))*

[**NOTE**: *No other information shall be highlighted.*]

A. If "*YES*," continue.
B. If "*NO*," state that the heading for _____ (specify heading) in the Nutrition Facts box fails to be highlighted as required by *21 CFR 101.9(d)(1)(iv).*

Q *Are the Amount Per Serving, the calorie statements, and each nutrient and its corresponding percent Daily Value separated by a hairline that is centered between the lines of text? (21 CFR 101.9(d)(1)(v))*
A. If "*YES*," continue.
B. If "*NO*," that the _____ (specify item that fails to comply) is not separated by a hairline as required by *21 CFR 101.9(d)(1)(v).*

Q *Does the heading "Nutrition Facts" in the Nutrition Facts box appear in a larger type size than all other print in the box and set the full width of the box? (21 CFR 101.9(d)(2))*

A. If "*YES*," continue.
B. If "*NO*," state that the heading "Nutrition Facts" is not declared _____ (specify deviation) as required by *21 CFR 101.9(d)(2).*

Q *Does the information on "serving size" immediately follow the heading "Nutrition Facts"? (21 CFR 101.9(d)(3))*
A. If "*YES*," continue.
B. If "*NO*," state that the serving size declaration does not immediately follow the heading "Nutrition Facts" in the Nutrition Facts box.

Q *Does the Nutrition Facts box contain a declaration of "Servings Per Container," (not required on single serving containers - 21 CFR 101.9(d)(3)(ii)), "Amount Per Serving" (separated by a bar - 21 CFR 101.9(d)(4)), information on calories (immediately following Amount Per Serving and declared on one line (21 CFR 101.9(d)(5)), and "% Daily Value" followed by an asterisk and separated by a bar (the term "Percent DV," or "% DV" may be substituted for "% Daily Value")? (21 CFR 101.9(d)(6))*

A. If "*YES*," continue.
B. If "*NO*," state that the _____ (specify the deviation) fails to be declared in the Nutrition Facts box as required by _____ (insert the reference for the specific requirement).

Q *Is each nutrient declared between the calorie and the vitamin and mineral sections of the Nutrition Facts box listed in a column and immediately followed by quantitative amounts by weight in terms of grams or milligrams with a corresponding declaration under the heading "% Daily Value," expressed to the nearest whole percent, except where exempted?*

A. If "*YES*," continue.
B. If "*NO*," state that the declaration for _____ (specific nutrient not so declared) fails to be declared in the Nutrition Facts box as required by *21 CFR 101.9(d)(7)(i) and (ii).*

Chapter VI

Q

Are the declarations for vitamins and minerals with numerical values separated in the Nutrition Facts box from other nutrients by a bar? *(21 CFR 101.9(d)(8))*

A. If "*YES*," continue.
B. If "*NO*," state that the declarations for vitamins and minerals in the Nutrition Facts box fail to be separated from other nutrients by a bar as required by *21 CFR 101.9(d)(8)*.

Q

Is the footnote, which is preceded by an asterisk, placed beneath the declaration of vitamins and minerals and separated by a hairline in the Nutrition Facts box? *(21 CFR 101.9(d)(9))*

A. If "*YES*," continue.
B. If "*NO*," state that the footnote fails to be _____ (specify the deviation, e.g., separated by a hairline in the Nutrition Facts box) as required by *21 CFR 101.9(d)(9)*.

Q

Is the nutrition information presented for two or more forms of the same food, "as packaged" and "as prepared" as provided for under 21 CFR 101.9(e)? *(See Illustration VI.E.4(c))*

A. If "*YES*," as in 1 and 2 below, continue.
    *   Information for each form shall be of equal prominence.
    *   All nutrients and values shall be listed under both columns.
B. If "*NO*," because the nutrition information for two or more forms is not declared in equal prominence, state that the nutrition information is not presented in equal prominence for the forms declared as required by *21 CFR 101.9(e)*.

[NOTE: If the product has separately packaged ingredients or foods, consists of an assortment of foods, or is a food to which other ingredients are added by the user; it must bear nutrition labeling in accordance with *21 CFR 101.9(h)*.] *(See Regulations VI.E.7)*

Q

Are the declarations for the % Daily Value in the Nutrition Facts box declared in accordance with the "Rounding Rule Table for Declaring Nutrients"? *(See Chart VI.E.8)*
A. If "*YES*," continue.
B. If "*NO*," state that the % Daily Value declarations for _____ and _____ (insert nutrients not properly rounded) are not properly rounded.

Q

Is the serving size declaration in the Nutrition Facts box declared in accordance with the "Rounding Rule Table for Serving Size"? *(See Chart VI.E.9)*

A. If "*YES*," continue.
B. If "*NO*," state that the serving size declaration in the Nutrition Facts box fails to be rounded to the appropriate increment.

Q

Are the substances and nutrients declared in accordance with applicable regulations?

A. If "*YES*," continue.
B. If "*NO*," state that the _____ (insert nutrient or substance that deviates from the requirements) declaration in the Nutrition Facts box fails to _____ (insert specific deviation) in accordance with _____ (insert reference to the specific requirement).

Chart VI.E.3
Format Styles for Nutrition Labeling

Chapter VI

Format Style/ Available Label Space	More Than 40 Square Inches	40 Or Less Square Inches	Less Than 12 Square Inches
Full vertical format with footnote	Yes	Yes	Yes
Full vertical format with no footnote	No	Yes	Yes
Simplified vertical format with no footnote	Yes	Yes	Yes
Tabular format with no footnote[1]	No	Yes	Yes
Tabular format with footnote[2]	Yes	Yes	Yes
Linear format[3]	No	Yes	Yes
Phone number or address for Obtaining nutrition information	No	No	Yes, if label bears no claims or other nutrition information

Notes: For discussion of "available label space," see August 18, 1993 *Federal Register* (58 FR 44075, comment 42) and *(21 CFR 101.9(j)(13))*.

1. "No footnote," as used in the above table, is to indicate that the footnote of *§101.9(d)(9)* is not needed. However, the statement "Percent Daily Value (DV) is based on a 2,000 calorie diet" is required on all labels, except on foods for children under four years of age.

2. The tabular format may be used only if there is not sufficient continuous space for the vertical format of the full, shortened, or simplified formats.

3. The linear format may be used only if the package size and shape cannot accommodate a column format.

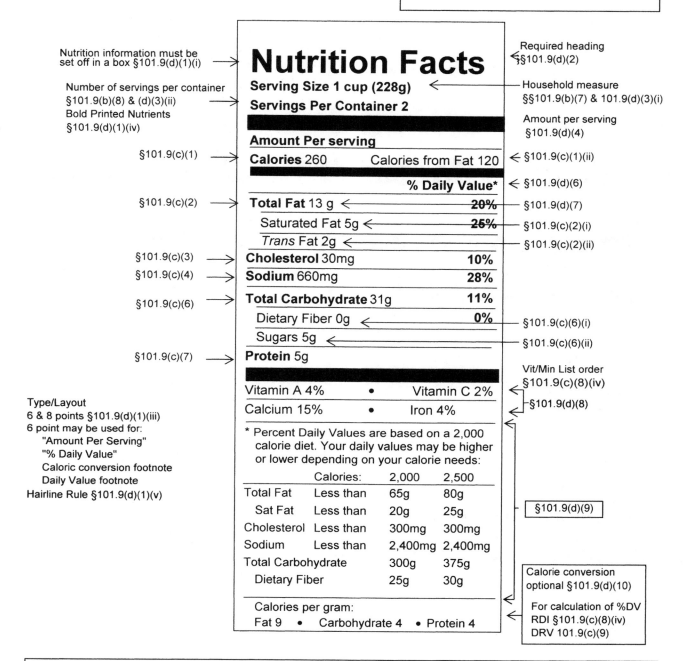

Nutrition information must be set off in a box §101.9(d)(1)(i)

Number of servings per container §101.9(b)(8) & (d)(3)(ii)
Bold Printed Nutrients §101.9(d)(1)(iv)

§101.9(c)(1)

§101.9(c)(2)

§101.9(c)(3)

§101.9(c)(4)

§101.9(c)(6)

§101.9(c)(7)

Type/Layout
6 & 8 points §101.9(d)(1)(iii)
6 point may be used for:
    "Amount Per Serving"
    "% Daily Value"
    Caloric conversion footnote
    Daily Value footnote
Hairline Rule §101.9(d)(1)(v)

Nutrition Facts
Serving Size 1 cup (228g)
Servings Per Container 2

Amount Per serving
Calories 260   Calories from Fat 120

                                    % Daily Value*

Total Fat 13 g   ~~20%~~
    Saturated Fat 5g   ~~25%~~
    Trans Fat 2g
Cholesterol 30mg   **10%**
Sodium 660mg   **28%**
Total Carbohydrate 31g   **11%**
    Dietary Fiber 0g   **0%**
    Sugars 5g
Protein 5g

Vitamin A 4%   •   Vitamin C 2%
Calcium 15%   •   Iron 4%

* Percent Daily Values are based on a 2,000 calorie diet. Your daily values may be higher or lower depending on your calorie needs:

	Calories:	2,000	2,500
Total Fat	Less than	65g	80g
Sat Fat	Less than	20g	25g
Cholesterol	Less than	300mg	300mg
Sodium	Less than	2,400mg	2,400mg
Total Carbohydrate		300g	375g
Dietary Fiber		25g	30g

Calories per gram:
Fat 9   •   Carbohydrate 4   •   Protein 4

Required heading §§101.9(d)(2)

Household measure §§101.9(b)(7) & 101.9(d)(3)(i)

Amount per serving §101.9(d)(4)

§101.9(c)(1)(ii)

§101.9(d)(6)

§101.9(d)(7)

§101.9(c)(2)(i)

§101.9(c)(2)(ii)

§101.9(c)(6)(i)

§101.9(c)(6)(ii)

Vit/Min List order §101.9(c)(8)(iv)

§101.9(d)(8)

§101.9(d)(9)

Calorie conversion optional §101.9(d)(10)

For calculation of %DV
RDI §101.9(c)(8)(iv)
DRV 101.9(c)(9)

Notes:

1. The terms "*Nutrition Facts*," "*Amount Per Serving*," "*% Daily Value*," and the names of all nutrients not required to be indented, (i.e., "*Calories*," "*Total Fat*," "*Cholesterol*," "*Sodium*," "*Total Carbohydrate*," and "*Protein*"), as well as the percentage amounts for the % DV, shall be highlighted by bold or extra bold type in the nutrition label. No other information shall be highlighted. (*21 CFR 101.9(d)(1)(iv)*)

2. The heading, "*Nutrition Facts*," shall be set in a type size larger than all other printed matter in the nutrition label and, unless impractical, be set the full width of the information provided in the box, except as permitted by *§101.9(d)(11)* (in cases where there is insufficient space on the label). (*21 CFR 101.9(d)(2)*)

Illustration VI.E.4(b)
Full Vertical Format with
Footnote to the Side

Chapter VI

Nutrition Facts

Serving Size 1 cup (228g)

Servings Per Container 2

Amount Per serving

Calories 260   Calories from Fat 120

% Daily Value*

Total Fat 13g	**20%**
Saturated Fat 5g	**25%**
Trans Fat 2g	
Cholesterol 30mg	**10%**
Sodium 660mg	**28%**
Total Carbohydrate 31g	**10%**
Dietary Fiber 0g	**0%**
Sugars 5g	
Protein 5g	

Vitamin A 4%	•	Vitamin C 2%
Calcium 15%	•	Iron 4%

* Percent Daily Values are based on a 2,000 calorie diet. Your daily values may be higher or lower depending on your calorie needs:

	Calories:	2,000	2,500
Total Fat	Less than	65g	80g
Sat Fat	Less than	20g	25g
Cholesterol	Less than	300mg	300mg
Sodium	Less than	2,400mg	2,400mg
Total Carbohydrate		300g	375g
Dietary Fiber		25g	30g

Calories per gram:

Fat 9   •   Carbohydrate 4   •   Protein 4

Notes:

Section *101.9(d)(11)* provides the following two provisions for breaking the vertical format of the full nutrition label:

1) Section *101.9(d)(11)(i)* permits the "%DV" footnote of section *101.9(d)(9)* to be moved anywhere directly to the right of the column when the space below the listing of vitamins and minerals is insufficient (as illustrated above).
2) Section *101.9(d)(9)(11)(ii)* permits all nutrients listed after "iron" and "%DV" footnote to be moved anywhere directly to the right of the column when the space below the listing of vitamins and mineral is insufficient.

No other breaks in the label are permitted.

Illustration VI.E.4(c)
Full Vertical Formats with Dual Declaration

As Packaged and As Prepared

Nutrition Facts

Serving Size $^1/_{12}$ package
(44g, about $^1/_4$ cup dry mix)
Servings Per Container 12

Amount Per Serving	Mix	Baked
Calories	190	280
Calories from Fat	45	140

	%Daily Value**	
Total Fat 5g*	8%	24%
Saturated Fat 2g	10%	13%
Trans Fat 1g		
Cholesterol 0mg	0%	23%
Sodium 300mg	13%	13%
Total Carbohydrate 34g	11%	11%
Dietary Fiber 0g	0%	0%
Sugars 18g		
Protein 2g		

Vitamin A	0%	0%
Vitamin C	0%	0%
Calcium	6%	8%
Iron	2%	4%

* Amount in mix

** Percent Daily Values are based on a 2,000 calorie diet. Your daily values may be higher or lower depending on your calorie needs.

	Calories	2,000	2,500
Total Fat	Less than	65g	80g
Sat Fat	Less than	20g	25g
Cholesterol	Less than	300mg	300mg
Sodium	Less than	2,400mg	2,400mg
Total Carbohydrate		300g	375g
Dietary Fiber		25g	30g

Calories Per Gram:
Fat 9 • Carbohydrate 4 • Protein 4

Nutrients Added by Combination of Foods

Nutrition Facts

Serving Size 1 cup (35g)
Servings Per Container 10

Amount Per Serving	Cereal	Cereal with 1/2 Cup Skim Milk
Calories	130	170
Calories from Fat	0	0

	%Daily Value**	
Total Fat 0g*	0%	0%
Saturated Fat 0g	0%	0%
Trans Fat 0g		
Cholesterol 0mg	0%	0%
Sodium 200mg	8%	11%
Total Carbohydrate 30g	10%	12%
Dietary Fiber 4g	16%	16%
Sugars 18g		
Protein 3g		

Vitamin A	25%	25%
Vitamin C	25%	25%
Calcium	0%	15%
Iron	10%	10%

* Amount in Cereal. One half cup skim milk contributes an additional 40 calories, 65mg sodium, 6g total carbohydrates (6g sugars) and 4g protein.

** Percent Daily Values are based on a 2,000 calorie diet. Your daily values may be higher or lower depending on your calorie needs:

	Calories	2,000	2,500
Total Fat	Less than	65g	80g
Sat Fat	Less than	20g	25g
Cholesterol	Less than	300mg	300mg
Sodium	Less than	2,400mg	2,400mg
Total Carbohydrate		300g	375g
Dietary Fiber		25g	30g

Calories Per Gram:
Fat 9 • Carbohydrate 4 • Protein 4

Adopted from FDA Guide to Nutritional Labeling and Educational Act (NLEA) Requirements, August 1994, revised December 30, 2005 and in 21 CFR Parts 100 through 169 (4/1/06 Edition)

Nutrition Facts/Datos Nutricional

Serving Size/Tamano por Racion 1 cup/1 taza (228g)
Servings Per Container/Raciones por Envase 2

Amount Per Serving/Cantidad por Racion

Calories/Calorias 260 Calories from Fat/Calorias de Grasa 120

	% Daily Value*/% Valor Diario*
Total Fat/Grasa Total 13g	**20%**
Saturated Fat/Grasa Saturada 5g	**25%**
Trans Fat 0g/Grasa *Trans* 0g	
Cholesterol/Colesterol 30mg	**10%**
Sodium/Sodio 660mg	**28%**
Total Carbohydrate/Carbohidrato Total 31g	**11%**
Dietary Fiber/Fibra Dietetica 0g	**0%**
Sugars/Azucares 5g	
Protein/Proteinas 5g	

Vitamin/Vitamina A 4%	•	Vitamin/Vitamina C 2%
Calcium/Calcio 15%	•	Iron/Hierro 4%

* Percent Daily Values are based on a 2,000 calorie diet. Your daily values may be higher or lower depending on your calorie needs:

* Los valores de los porcentajes Diarios estan basado en una dieta de 2,000 calorias. Sus valores diarios pueden ser mayor o menor dependiendo de sus necesidades caloricas:

		Calories/Calorias:	2,000	2,500
Total Fat/Grasa Total	Less than/Menos de		65g	80g
Sat Fat/Grasa Saturada	Less than/Menos de		20g	25g
Cholesterol/Colesterol	Less than/Menos de		300mg	300mg
Sodium/Sodio	Less than/Menos de		2,400mg	2,400mg
Total Carbohydrate/Carbohidrato Total			300g	375g
Dietary Fiber/Fibra dietetica			25g	30g

When nutrition labeling must appear in a second language, in accordance with *§101.15(c)(2)*, the nutrition information may be presented in a separate nutrition facts panel for each language or as shown in this example with the second language following the English. Numeric characters that are identical for both languages do not need to be repeated. All required nutrition information must be included in both languages.

Adopted from FDA Guide to Nutritional Labeling and Educational Act (NLEA)
Requirements, August 1994, revised December 30, 2005 and in 21 CFR Parts 100
through 169 (4/1/06 Edition)

Illustration VI.E.4(e)
Full Vertical Format with
Aggregate Declaration

Quantitative amount and % DV must
be listed in separate columns under
the name of each food

Name of each food identified in separate columns
21 CFR101.9(d)(13)(ii)

Nutrition Facts

	Wheat Squares Sweetened		Corn Flakes Not Sweetened		Mixed Grain Flakes Sweetened	
Serving Size 1 Box	(35g)		(19g)		(27g)	
Servings Per Container	1		1		1	
Amount Per Serving						
Calories	130		70		100	
Calories from fat	0		0		0	
	%Daily Value*		**%Daily Value***		**%Daily Value***	
Total Fat	0g	**0%**	0g	**0%**	0g	**0%**
Saturated Fat	0g	**0%**	0g	**0%**	0g	**0%**
Trans Fat	0g		0g		0g	
Cholesterol	0g	**0%**	0g	**0%**	0g	**0%**
Sodium	0g	**0%**	200mg	**8%**	120g	**5%**
Potassium	125g	**4%**	25mg	**1%**	30g	**1%**
Total Carbohydrate	29g	**10%**	17g	**6%**	24g	**8%**
Dietary Fiber	3g	**12%**	1g	**4%**	1g	**4%**
Sugars	8g		6g		13g	
Protein	4g		1g		1g	
Vitamin A	0%		10%		10%	
Vitamin C	0%		15%		90%	
Calcium	0%		0%		0%	
Iron	10%		6%		20%	
Thiamin	30%		15%		20%	
Riboflavin	30%		15%		20%	
Niacin	30%		15%		20%	
Vitamin B$_6$	30%		15%		20%	

* Percent Daily Values are based on a 2,000 calorie diet. Your daily values may be higher or lower depending on your calorie needs:

	Calories:	2,000	2,500
Total Fat	Less than	65g	80g
Sat Fat	Less than	20g	25g
Cholesterol	Less than	300mg	300mg
Sodium	Less than	2,400mg	2,400mg
Potassium	Less than	3,500mg	3,500mg
Total Carbohydrate		300g	375g

Calories per gram:

Fat 9 . Carbohydrates 4 . Protein 4

Vitamin/Mineral declaration in vertical
string, *21 CFR 101.9(d)(8)*

If potassium is declared among the nutrients of the middle section of the Nutrition Facts panel, *21 CFR 101.9 (d)(7)(i)*, potassium must also be listed under sodium in the Daily Value footnote, *21 CFR 101.9(d)(9)(iii)*.

Adopted from FDA Guide to Nutritional Labeling and Educational Act (NLEA) Requirements, August 1994, revised December 30, 2005 and in 21 CFR Parts 100 through 169 (4/1/06 Edition)

75

Illustration VI.E.4(f)
Simplified Vertical Formats

Chapter VI

This simplified format may be used if eight (8) or more of the following nutrients are present in *"insignificant amounts"*: calories, total fat, saturated fat, *trans* fat, cholesterol, sodium, total carbohydrate, dietary fiber, sugars, protein, vitamin A, vitamin C, calcium, and iron. Foods meeting this criterion may use the simplified format, regardless of package size.

21 CFR 101.9(f)(1) defines *"insignificant amount"* as the amount that allows a declaration of zero in nutrition labeling, except that for total carbohydrate, dietary fiber, and protein it shall be the amount that allows a declaration of *"less than 1 gram."*

There are five core nutrients (calories, total fat, sodium, total carbohydrate, and protein) that must be listed regardless of the quantity present in the food. Also any of the nutrients listed in *21 CFR 101.9(f)* that are present in more than insignificant amounts must also be listed. Label sample A shows the simplified format. The nutrients listed in sample B include core nutrients as well as nutrients that were included either voluntarily or because of the presence of a claim. When a product that qualifies to use the simplified format uses such format on its label or labeling, the statement "Not a significant source of _____" (with the blank filled with the name(s) of any nutrient(s) identified in section *101.9(f)* and calories from fat that are present in insignificant amounts) must be included at the bottom of the nutrition label.

Sample A (Soft Drink)

Nutrition Facts
Serving Size 1 can (360 ml)

Amount Per Serving

Calories 140

	% Daily Value*
Total Fat 0g	**0%**
Sodium 20mg	**1%**
Total Carbohydrate 36g	**12%**
Sugars 36g	
Protein 0g	

* Percent Daily Values are based on a 2,000 calorie diet.

Saturated, poly-, and monounsaturated fats
21 CFR 101.9(c)(2)(i), (ii) and (iii).

Note: When nutrients naturally present in the food are voluntarily declared and /or when products are enriched or when claims are made about a nutrient, additional requirements apply *21 CFR 101.9(f)(4).*

§101.9(f)(5) - required statement in lieu of footnote.

Sample B (Vegetable Oil)

Nutrition Facts
Serving Size 1 Tbsp (14g)
Servings Per Container 64

Amount Per Serving

Calories 130   Calories from Fat 130

	% Daily Value*
Total Fat 14g	**22%**
Saturated Fat 2g	**10%**
Trans Fat	
Polyunsaturated Fat 4g	
Monounsaturated Fat 8g	
Sodium 0mg	**0%**
Total Carbohydrate 0g	**0%**
Protein 0g	

Not a significant source of cholesterol, dietary fiber, sugars, vitamin A, vitamin C, calcium and iron.

* Percent Daily Values are based on a 2,000 calorie diet.

Adopted from FDA Guide to Nutritional Labeling and Educational Act (NLEA) Requirements,
August 1994, revised December 30, 2005 and in 21 CFR Parts 100 through 169 (4/1/06 Edition)

(Permitted on any size package)

Nutrition Facts

Serving Size 1 cup (245g)
Servings Per Container 2

Amount Per Serving

Calories 60   Calories from Fat 10

%Daily Value*

Total Fat 1g	**2%**
Sodium 800mg	**33%**
Total Carbohydrate 10g	**3%**
Dietary Fiber 1g	**4%**
Protein 2g	

Vitamin A 20% • Vitamin C 4% • Iron 2%

Not a significant source of saturated fat, trans fat, cholesterol, sugars, and calcium.

* Percent Daily Values are based on a 2,000 calorie diet. Your daily values may be higher or lower depending on your calorie needs:

	Calories:	2,000	2,500
Total Fat	Less than	65g	80g
Sat Fat	Less than	20g	25g
Cholesterol	Less than	300mg	300mg
Sodium	Less than	2,400mg	2,400mg
Total Carbohydrate		300g	375g
Dietary Fiber		25g	30

"Insignificant" is defined in §101.9(f)(1)

For CFR reference, see individual nutrient listing in *21 CFR 101.9(c)*.

The shortened format allows nutrients that are present at insignificant levels to be listed following the statement *"Not a significant source of _____"*. One or more of the nutrients that qualify for such a declaration are listed in the order in which they would have been listed in the regular format, e.g., *"Not a significant source of calories from fat, saturated fat, trans fat, cholesterol, dietary fiber, sugars, vitamin A, vitamin C, calcium and iron"*. This statement may be printed in 6 point type.

The footnote can be used with any format, to list one or more of the following nutrients:

Calories from fat: when the food contains less than 5 calories from fat. *(21 CFR 101.9(c)(1)(ii))*

Saturated fat and Trans Fat: when the food contains less than 0.5 grams of total fat per serving and if no claims are made about fat, fatty acids or cholesterol content, and if no claims are made about calories from fat.
(21 CFR 101.9(c)(2)(i)) and (ii)

Cholesterol: when the food contains less than 2 milligrams cholesterol per serving, and makes no claim about fat, fatty acids or cholesterol.
(21 CFR 101.9(c)(3))

Dietary fiber: when a serving contains less than 1 gram of dietary fiber. *(21 CFR 101.9(c)(6)(i))*

Sugars: when a serving contains less than 1 gram of sugar, and no claims are made about sweeteners, sugars, or sugar alcohol content.
(21 CFR 101.9(c)(6)(ii))

Vitamins/Minerals: when a serving contains less than 2% of the RDI. *(21 CFR 101.9(c)(8)(iii))*

Adopted from FDA Guide to Nutritional Labeling and Education Act (NLEA)
Requirements, August 1994, and 21 CFR Parts 100 through 169 (4/1/06 Edition)

77

Illustration VI.E.4(h)
Full Tabular Format

Chapter VI

Nutrition Facts	Amount/serving	%Daily Value*	Amount/serving	%Daily Value*	* Percent Daily Values are based on a 2,000 calorie diet. Your daily values may be higher or lower depending on your calorie needs:

Nutrition Facts Serving Size 2 slices (56g) Servings Per Container 10	**Total Fat** 1.5g	**2%**	**Total Carbohydrate** 26g	**9%**	Calories: 2,000 / 2,500

Nutrition Facts

Serving Size 2 slices (56g)
Servings Per Container 10

Calories 140
   Calories from Fat 15

Amount/serving	%Daily Value*	Amount/serving	%Daily Value*
Total Fat 1.5g	**2%**	**Total Carbohydrate** 26g	**9%**
Saturated Fat 0g	**0%**	Dietary Fiber 2g	**8%**
Trans Fat 0.5g		Sugars 1g	
Cholesterol 0mg	**0%**	**Protein** 4g	
Sodium 280mg	**12%**		

Vitamin A 0% • Vitamin C 0% •   Calcium 6% • Iron 6%
Thiamin 15% • Riboflavin 8% •   Niacin 10% •

		2,000	2,500
Total Fat	Less than	65g	80g
Sat Fat	Less than	20g	25g
Cholesterol	Less than	300mg	300mg
Sodium	Less than	2,400mg	2,400mg
Total Carbohydrate		300g	375g
Dietary Fiber		25g	30g

Calories per gram:
Fat 9 • Carbohydrate 4 • Protein 4

Conditions of Use:

The technical corrections of August 18, 1993, provided for using a tabular format for packages with more than 40 square inches of available space only when there is not sufficient vertical space to accommodate the full format (approximately 3 inches) up to and including the declaration of iron *(§101.9(d)(11)(iii))*. The footnote is required.

Illustration VI.E.4(i)
Tabular Format for Small Packages

Tabular display for packages having 40 or less square inches of total surface area available to bear labeling.

Nutrition Facts

Serving Size 1/3 cup (56g)
Servings about 3
Calories 90
Fat Cal. 20

* Percent Daily Values (DV) are based on a 2,000 calorie diet.

Amount/serving	%DV*	Amount/serving	%DV*
Total Fat 2g	**3%**	**Total Carb.** 0g	**0%**
Sat. Fat 1g	**5%**	Fiber 0g	**0%**
Trans Fat 0.5		Sugars 0g	
Cholest. 10mg	**3%**	**Protein** 17g	
Sodium 200mg	**8%**		

Vitamin A 0% • Vitamin C 0% • Calcium 0% • Iron 6%

Vitamins and minerals information can be arranged horizontally
(21 CFR 101.9(d)(8))

The footnote may be omitted
(21 CFR 101.9(j)(13)(ii)(C))

This Tabular display may be used only if the packages cannot accommodate a column display on any panel.

Special Provisions:

Abbreviations may be used to list nutrients on packages with 40 or less square inches of available label space.

Abbreviations, listed below, were revised in the August 18, 1993 Federal Register *[21 CFR 101.9(j)(13)(ii)(B)]*.

Serving size	Serv size	Soluble fiber	Sol fiber
Servings per container	Servings	Insoluble fiber	Insol fiber
Calories from fat	Fat cal	Sugar alcohol	Sugar alc
Calories from saturated fat	Sat fat cal	Other	
Saturated fat	Sat fat	Carbohydrate	Other carb
Monounsaturated fat	Monounsat fat		
Polyunsaturated fat	Polyunsat fat		
Cholesterol	Cholest		
Total carbohydrate	Total carb		
Dietary fiber	Fiber		

Illustration VI.E.4(j)
Simplified Tabular Format

Chapter VI

Nutrition Facts

Serving Size 1 oz.
(28g/about 12 slices)
Servings Per Container 16
Calories 20
* Percent Daily Values are based
  on a 2,000 calorie diet

Amount /serving	% Daily Value*
Total Fat 0g	0%
Sodium 190mg	8%
Total Carbohydrate 5g	2%
Sugars 5g	
Protein 0g	

The simplified tabular format may be used on any size package of food which qualifies for the simplified format when there is not sufficient vertical space to accommodate the vertical simplified format.

Adopted from ORA, U.S. FD, August 1994 and 21 CFR Parts 100
through 169 (4/1/06 Edition)

Illustration VI.E.4(k)
Linear Formats

The linear display may be used on packages with 40 or less square inches of available space only if the label will not accommodate the tabular display. *(21 CFR 101.9(j)(13)(ii)(A))*

Full Linear Format

Nutrition Facts
Serv. Size: 1 package, Amount per Serving: **Calories** 45, Fat Cal. 10, **Total fat** 1g (2% DV), Sat. Fat 0.5g (3% DV), Trans Fat 0.5g, **Cholest.** 0mg (0% DV), **Sodium** 50mg (2% DV), **Total Carb.** 8g (3% DV), Fiber 1g (4% DV), Sugars 4g, **Protein** 1g, Vitamin A (8% DV), Vitamin C (8% DV), Calcium (0% DV). Percent Daily Values (DV) are based on a 2,000 calorie diet.

Shortened Linear Format

Nutrition Facts
Serv. Size: 1 package, Amount per Serving: **Calories** 40, **Total fat** 0g (0% DV), **Sodium** 50mg (2% DV), **Total Carb.** 8g (3% DV), Fiber 1g (4% DV), Sugars 4g, **Protein** 1g, Vitamin A (8% DV), Vitamin C (8% DV), Iron (2% DV). Not a significant source of calories from fat, saturated fat, trans fat, cholesterol, or calcium. Percent Daily Values (DV) are based on a 2,000 calorie diet.

Simplified Linear Format

Nutrition Facts
Serv. Size: 3 pieces (6g), Servings: 4, Amount Per Serving: **Calories** 20, **Total fat** 0g (0% DV). **Sodium** 20 mg (1% DV), **Total Carb.** 5g (2% DV), Sugars 5g, **Protein** 0g, Percent Daily Values (DV) are based on a 2,000 calorie diet.

Bolding is required only for "*Nutrition Facts*". Bolding is voluntary for nutrient names, calories, total fat, cholesterol, sodium, total carbohydrate, and protein. *(21 CFR 101.9(j)(13)(ii)(A)(2) - August 18, 1993)*

Footnote is not required. *(21 CFR 101.9(j)(13)(ii)(C))*

Abbreviations may be used. Abbreviations were revised in the August 18, 1993, Federal Register. (See *21 CFR 101.9(j)(13)(ii)(B))*

Nutrition Facts

Serving Size 1 jar (140g)

Amount Per Serving

Calories 110

Total Fat	**0g**
Sodium	**10mg**
Total Carbohydrate 27g	
Dietary Fiber	**4g**
Sugars	**18g**
Protein	**0g**

%Daily Value

Protein 0%	•	Vitamin A 6%
Vitamin C 45%	•	Calcium 2%
Iron 2%		

Nutrients that cannot be listed according to *21 CFR101.9(j)(5)(i)*.
- Calories from fat
- Calories from saturated fat
- Saturated fat
- Polyunsaturated fat
- Monounsaturated fat
- Cholesterol

% DV prohibited for total fat, saturated fat, cholesterol, sodium, potassium, total carbohydrate, and dietary fiber. *(21 CFR 101.9(j)(5)(ii)(A))*

Two column format: nutrient names and amounts by weight. *(21 CFR 101.9(j)(5)(ii)(B))*

% DV heading required. *(21 CFR 101.9(j)(5)(ii)(C))*

% DV heading required for vitamins, minerals, and protein* *(21 CFR 101.9(j)(5)(ii)(D))*

Footnote is prohibited. *(21 CFR 101.9(j)(5)(ii)(E))*

*21 CFR 101.9 (c)(7) provides special provisions for protein.

Infants - % DV for protein is prohibited on infant foods if the protein efficiency ratio is less than 40% of the reference standard. Such foods must omit the % DV and use the statement "Not a significant source of protein" adjacent to the declaration of protein content.

Children between 1 and 4 - If protein digestibility-corrected amino acid score is less than 40%, either the % DV or the statement "Not a significant source of protein" must be used.

Adopted from ORA, U.S. FD, August 1994, revised December 30, 2005

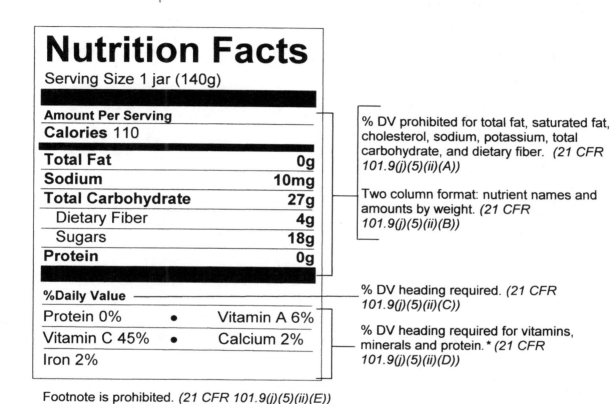

Nutrition Facts

Serving Size 1 jar (140g)

Amount Per Serving

Calories 110

Total Fat	**0g**
Sodium	**10mg**
Total Carbohydrate	**27g**
Dietary Fiber	**4g**
Sugars	**18g**
Protein	**0g**

%Daily Value

Protein 0%	●	Vitamin A 6%
Vitamin C 45%	●	Calcium 2%
Iron 2%		

% DV prohibited for total fat, saturated fat, cholesterol, sodium, potassium, total carbohydrate, and dietary fiber. *(21 CFR 101.9(j)(5)(ii)(A))*

Two column format: nutrient names and amounts by weight. *(21 CFR 101.9(j)(5)(ii)(B))*

% DV heading required. *(21 CFR 101.9(j)(5)(ii)(C))*

% DV heading required for vitamins, minerals and protein.* *(21 CFR 101.9(j)(5)(ii)(D))*

Footnote is prohibited. *(21 CFR 101.9(j)(5)(ii)(E))*

21 CFR 101.9(c)(7) provides special provisions for protein.

Children between 1 and 4 - If protein digestibility - corrected amino acid score is less than 40%, either the % DV or the statement "Not a significant source of protein" must be used.

Chart VI.E.5
Reference Amounts for Infant
and Toddler Foods
(21CFR 101.12)

Table 1
Reference Amounts Customarily Consumed Per Eating Occasion:
Infant and Toddler Foods[1,2,3,4]

Product Category	Reference Amount	Label Statement[5]
Cereals, dry instant	15g********	_____cup (____g)
Cereals, prepared, ready-to-serve	110g*******	_____cup(s) (____g)
Other cereal and grain products, dry ready-to-eat, e.g. ready-to-eat cereals, cookies, teething biscuits, and toasts	7g for infants and 20g for toddlers for ready-to-eat cereals; 7g for all others	_____cup(s)(____g) for ready-to-eat cereals; _____piece(s) (____g) for others
Dinners, desserts, fruits, vegetables or soups dry mix	15g********	_____tbsp(s) (____); _____cup(s) (____g)
Dinners, desserts, fruits, vegetables or soups ready-to-serve, junior type	110g*******	_____cup(s) (____g); _____cup(s) (____mL)
Dinners, desserts, fruits, vegetables or soups ready-to-serve, strained type	60g*******	_____cup(s) (____g); _____cup(s) (____mL)
Dinners, stews or soups for toddlers, ready-to-serve	170g*******	_____cup(s) (____g); _____cup(s) (____mL)
Fruits for toddlers, ready-to-serve	125g*******	_____cup(s) (____g)
Vegetables for toddlers, ready-to-serve	70g*******	_____cup(s) (____g)
Eggs/egg yolks, ready-to-serve	55g*******	_____cup(s) (____g)
Juices, all varieties	120mL***	4 fl oz (120mL)

[1] These values represent the amount of food customarily consumed per eating occasion and were primarily derived from the 1977-1978 and the 1987-1988 Nationwide Food Consumption Surveys, conducted by the U.S. Department of Agriculture.

[2] Unless otherwise noted in the Reference Amount column, the reference amounts are for the ready-to-serve or almost ready-to-serve form of the product (e.g., heat and serve, brown and serve). If not listed separately, the reference amount of the unprepared form (e.g., dry cereal) is the amount required to make the reference amount of the prepared form. Prepared means prepared for consumption (e.g., cooked).

[3] Manufacturers are required to convert the reference amount to the label serving size in a household measure most appropriate to their specific product using the procedures in *21 CFR 101.9(b)*.

[4] Copies of the list of products for each product category are available from the Office of Nutritional Products, Labeling and Dietary Supplements (HFS-810), Center for Food Safety and Applied Nutrition, Food and Drug Administration, 200 C St. SW, Washington, DC 20204.

[5] The label statements are meant to provide guidance to manufacturers on the presentation of serving size information on the label, but they are not required. The term "piece" is used as a generic description of a discrete unit. Manufacturers should use the description of a unit that is most appropriate for the specific product (e.g., sandwich for sandwiches, cookie for cookies, and bar for frozen novelties).

Adopted from 21 CFR Parts 100 through 169 (4/1/06 Edition)

Chart VI.E.6
Reference Amounts for the
General Food Supply
(21CFR 101.12)

Table 2
Reference Amounts Customarily Consumed Per Eating Occasion: General Food Supply[1,2,3,4]

Product Category	Reference Amount	Label Statement[5]
Bakery Products:	*********************	
Biscuits, croissants, bagels, tortillas, soft bread sticks, soft pretzels, corn bread, hush puppies	55 g*****************	____piece(s) (____g)
Breads (excluding sweet quick type), rolls	50 g*****************	____piece(s) (__g) for sliced bread and distinct pieces (e.g., rolls); 2 oz (56g/____inch slice) for unsliced bread
Bread sticks - see crackers	*********************	
Toaster pastries - see coffee cakes	*********************	
Brownies	40 g*****************	____piece(s) (____g) for distinct pieces; fractional slice (____g) for bulk
Cakes, heavy weight (cheese cake; pineapple upside-down cake; fruit, nut and vegetable cakes with more than or equal to 35 percent of the finished weight as fruit, nuts, or vegetables or any of these combined)[6]	125 g***************	____piece(s) (____g) for distinct pieces (e.g., sliced or individually packaged products); fractional slice (____g) for large discrete units
Cakes, medium weight (chemically leavened cake with or without icing or filling except those classified as light weight cake; fruit, nut, and vegetable cake with less than 35 percent of the finished weight as fruit, nuts, or vegetables or any of these combined; light weight cake with icing; Boston cream pie; cupcake; eclair; cream puff)[7]	80 g*****************	____piece(s) (____g) for distinct pieces (e.g., cupcakes); ____fractional slice (____g) for large discrete units
Cakes, light weight (angel food, chiffon, or sponge cake without icing or filling)[8]	55 g*****************	____piece(s) (____g) for distinct pieces (e.g., sliced or individually packaged products); ____ fractional slice (____g)for large discrete units

Chart VI.E.6
(Continued)

Product Category	Reference Amount	Label Statement[5]
Bakery Products: *(Continued)*		
Coffee cakes, crumb cakes, doughnuts, Danish, sweet rolls, sweet quick type breads, muffins, toaster pastries	55 g*****************	____piece(s) (____g) for sliced bread and distinct pieces (e.g., doughnut); 2 oz (56 g/visual unit of measure) for bulk products (e.g., unsliced bread)
Cookies	30 g****************	____piece(s) (____g)
Crackers that are usually not used as snack, melba toast, hard bread sticks, ice cream cones[9]	15 g****************	____piece(s) (____g)
Crackers that are usually used as snacks	30 g****************	____piece(s) (____g)
Croutons	7 g******************	____tbsp(s) (____g); ____cup(s) (____g);
French toast, pancakes, variety mixes	110 g prepared for French toast and pancakes; 40g dry mix for variety mixes	____pieces (____g) for large pieces
Grain-based bars with or without filling or coating, e.g., breakfast bars, granola bars, rice cereal bars	40 g****************	___piece(s) (____g); ____cup(s) (____g) for dry mix
Ice cream cones - see crackers	**********************	
Pies, cobblers, fruit crisps, turnovers, other pastries	125 g**************	____piece(s) (____g) for distinct pieces; ____fractional slice (____g) for large discrete units
Pie crust	1/6 of 8 inch crust 1/8 of 9 inch crust	1/6 of 8 inch crust (____g); 1/8 of 9 inch crust (____g)
Pizza crust	55 g****************	____fractional slice (____g)

Adopted from 21 CFR Parts 100 through 169 (4/1/06 Edition)

Chart VI.E.6
(Continued)

Product Category	Reference Amount	Label Statement[5]
Bakery Products: *(Continued)*	**********************	
Taco shell, hard	30 g*****************	____shell(s) (____g)
Waffles	85 g*****************	____piece(s) (____g)
Beverages:	**********************	
Carbonated and noncarbonated beverages, wine coolers, water	240 mL*************	8 fl oz (240mL)
Coffee or tea, flavored and sweetened	240 mL prepared**	8 fl oz (240mL)
Cereal and Other Grain Products:	**********************	
Breakfast cereals (hot cereal type), hominy grits	1 cup prepared, 40 g plain dry cereal, 55 g flavored, sweetened dry cereal	____cup(s) (____g)
Breakfast cereals, ready-to-eat, weighing less than 20g per cup, e.g., plain puffed cereal grains	15 g*****************	____cup(s) (____g)
Breakfast cereals, ready-to-eat, weighing 20 g or more but less than 43 g per cup; high fiber cereals containing 28 g or more of fiber per 100 g	30 g*****************	____cup(s) (____g)
Breakfast cereals, ready-to-eat, weighing 43 g or more per cup; biscuit types.	55 g*****************	____piece(s) (____g) for large distinct pieces (e.g., biscuit type; ____cup(s) (____g) for all others
Bran or wheat germ	15 g*****************	____tbsp(s) (____g); ____cup(s) (____g)
Flours or cornmeal	30 g*****************	____tbsp(s) (____g); ____cup(s) (____g)
Grains, e.g., rice, barley, plain	140 g prepared; 45 g dry	____cup(s) (____g)
	**********************	

Chart VI.E.6
(Continued)

Chapter VI

Product Category	Reference Amount	Label Statement[5]
Cereal and Other Grain Products: *(Continued)*		
Pastas, plain	140 g prepared; 55 g dry***********	____cup(s) (____g); ____piece(s) (____g) for large pieces (e.g., large shells or lasagna noodles) or 2 oz (56 g/visual unit of measure) for dry bulk products (e.g., spaghetti)
Pastas, dry, ready-to-eat, e.g., fried canned chow mein noodles	25 g*****************	____cup(s) (____)g
Starches, e.g., cornstarch, potato starch, tapioca, etc.	10 g*****************	____tbsp (____g)
Stuffing	100 g***************	____cup(s) (____)g
Dairy Products and Substitutes:	*********************	
Cheese, cottage	110 g**************	____cup(s) (____)g
Cheese used primarily as ingredients, e.g., dry cottage cheese, ricotta cheese	55 g*****************	____cup(s) (____)g
Cheese, grated hard, e.g., Parmesan, Romano	5 g*****************	____tbsp (____g)
Cheese, all others except those listed as separate categories - includes cream cheese and cheese spread	30 g*****************	____piece(s) (____g) for distinct pieces; ____tbsp(s) (____g) for cream cheese and cheese spread; 1 oz (28 g/visual unit of measure) for bulk
Cheese sauce - see sauce category	*********************	
Cream or cream substitutes, fluid	15 mL***************	1 tbsp (15 mL)
Cream or cream substitutes, powder	2 g*****************	____tsp (____g)
Cream, half and half	30 mL***************	2 tbsp (30 mL)
Eggnog	120 mL*************	1/2 cup (120 mL); 4 fl oz (120 mL)

Adopted from 21 CFR Parts 100 through 169 (4/1/06 Edition)

Chart VI.E.6
(Continued)

Product Category	Reference Amount	Label Statement[5]
Dairy Products and Substitutes: *(Continued)*	*********************	
Milk, condensed, undiluted	30 mL**************	2 tsp (30 mL)
Milk, evaporated, undiluted	30 mL**************	2 tsp (30 mL)
Milk, milk-based drinks, e.g., instant breakfast, meal replacement, cocoa	240 mL************	1 cup (240 mL); 8 fl oz (240 mL)
Shakes or shake substitutes, e.g., dairy shake mixes, fruit frost mixes	240 mL************	1 cup (240 mL); 8 fl oz (240 mL)
Sour cream	30 g***************	____tbsp (____g)
Yogurt	225 g*************	____cup(s) (____g)
Desserts:	*********************	
Ice cream, ice milk, frozen yogurt, sherbet: all types, bulk and novelties (e.g., bars, sandwiches, cones)	1/2 cup - includes the volume for coating and wafers for the novelty type varieties	____piece(s) (____g) for individually wrapped or packaged products; 1/2 cup (____g) for others
Frozen flavored and sweetened ice and pops, frozen fruit juices: all types, bulk and novelties (e.g., bars, cups)	85 g****************	____piece(s) (____g) for individually wrapped or packaged products; ____cup(s) (____g) for others
Sundae	1 cup**************	1 cup (____g)
Custards, gelatin or pudding	1/2 cup************	____piece(s) (____) for distinct unit (e.g., individually packaged products); 1/2 cup (____g) for bulk
Dessert Toppings and Fillings:	*********************	
Cake frostings or icings	35 g****************	____tbsp(s) (____g)
Other dessert toppings, e.g., fruits, syrup spreads, marshmallow cream, nuts, dairy and nondairy whipped toppings	2 tbsp**************	2 tbsp (____g); 2 tbsp (30 mL))

Chart VI.E.6
(Continued)

Product Category	Reference Amount	Label Statement[5]
Dessert Toppings and Fillings: *(Continued)*	**********************	
Pie filling	85 g*****************	____cup(s) (____g)
Egg and Egg Substitutes:	**********************	
Egg mixture e.g., egg foo yung scrambled eggs, omelets	110 g**************	____piece(s) (____g) for discrete pieces; ____cup(s) (____g)
Eggs (all sizes)[9]	50 g****************	1 large, medium, etc. (____g)
Egg substitutes	An amount to make 1 large (50 g) egg	____cup(s) (____g); ____cup(s) (____mL)
Fats and Oils:	**********************	
Butter, margarine, oil, shortening	1 tbsp**************	1 tbsp (____g); 1 tbsp (15 mL)
Butter replacement, powder	2 g*****************	____tsp(s) (____g)
Dressings for salads	30 g****************	____tbsp (____g); ____tbsp (____mL)
Mayonnaise, sandwich spreads, mayonnaise type dressings	15 g****************	____tbsp (____g)
Spray types	0.25 g**************	About ____seconds spray) (____g)
Fish, Shellfish, Game Meats[10], and Meat or Poultry Substitutes:	**********************	
Bacon substitutes, canned anchovies,[11] anchovy pastes, caviar	15 g****************	____piece(s) (____g) for discrete pieces; tbsp(s) (____g) for others
Dried, e.g., jerky	30 g****************	____piece(s) (____g)
Entrees with sauce, e.g., fish with cream sauce, shrimp with lobster sauce	140 g cooked******	____cup(s) (____g); 5 oz (140 g/visual unit of measure) if not measurable by cup

Adopted from 21 CFR Parts 100 through 169 (4/1/06 Edition)

Chart VI.E.6
(Continued)

Product Category	Reference Amount	Label Statement[5]
Fish, Shellfish, Game Meats[10], and Meat or Poultry Substitutes: *(Continued)*	**********************	
Entrees without sauce, e.g., plain or fried fish and shellfish, fish and shellfish cake	85 g cooked; 110 g uncooked[12]***********	____piece(s) (____g) for discrete pieces; ____ cup(s) (____g); ____oz (____g/visual unit of measure) if not measurable by cup[13]
Fish, shellfish or game meat, canned[11]	55 g****************	____pieces(s) (____g) for discrete pieces; ____cup(s) (____g); 2oz (56 g/ ____cup) for products that are difficult to measure the g weight of cup measure (e.g., tuna); 2 oz (56 g/____pieces) for products that naturally vary in size (e.g., sardines)
Substitute for luncheon meat, meat spreads, Canadian bacon, sausages and frankfurters	55 g****************	____pieces(s) (____g for distinct pieces (e.g., slices, links); ____cup(s) (____g); 2 oz (56 g/ visual unit of measure) for nondiscrete bulk product
Smoked or pickled fish, shellfish, or game meat[10]; fish or shellfish spread	55 g****************	____piece(s) (____g for distinct pieces (e.g., slices, links) or ____cup(s) (____g; 2 oz (56 g/visual unit of measure) for nondiscrete bulk product
Substitute for bacon bits - see miscellaneous category	**********************	
Fruits and Fruit Juices:	**********************	
Candied or pickled[11]	30 g****************	____piece(s) (____g)
Dehydrated fruits - see snacks category	**********************	

Chart VI.E.6
(Continued)

Chapter VI

Product Category	Reference Amount	Label Statement[5]
Fruits and Fruit Juices: *(Continued)*	**********************	
Dried	40 g*****************	____piece(s) (____g) for large pieces (e.g., dates, figs, prunes); ____cup(s) (____g) for small pieces (e.g., raisins)
Fruits for garnish or flavor, e.g., maraschino cherries[11]	4 g******************	1 cherry (____g)
Fruit relishes, e.g., cranberry sauce, cranberry relish	70 g****************	____cup(s) (____g)
Fruits used primarily as ingredients, avocado	30 g****************	See footnote 13
Fruits used primarily as ingredients, others (cranberries, lemon, lime)	55 g****************	____pieces(s) (____g for large fruits; ____cup(s) (____g) for small fruits measurable by cup
Watermelon	280 g**************	See footnote 13
All other fruits (except those listed as separate categories), fresh, canned or frozen	140 g**************	____piece(s) (____g) for large pieces (e.g., strawberries, prunes, apricots, etc.); ____cup(s) (____g) for small pieces (e.g., blueberries, raspberries, etc.)
Juices, nectars, fruit drinks	240 mL*************	8 fl oz (240 mL)
Juices used as ingredients, e.g., lemon juice, lime juice	5 mL***************	1 tsp (5 mL)
Legumes:	**********************	
Bean cake (tofu)[11], tempeh	85 g****************	____piece(s) (____g) for discrete pieces; 3oz (84 g/visual unit of measure for bulk products
Beans, plain or in sauce	130 g for beans in sauce or canned in liquid and refried beans prepared; 90 g for others prepared; 35 g dry.	____cup (____g)

Chart VI.E.6
(Continued)

Product Category	Reference Amount	Label Statement[5]
Miscellaneous Category:	**********************	
Baking powder, baking soda, pectin	0.6 g******************	_____tsp (_____g)
Baking decorations, e.g., colored sugars and sprinkles for cookies, cake decorations	1 tsp or 4 g if not measurable by teaspoon	_____piece(s) (_____g) for discrete pieces; 1 tsp (_____g)
Batter mixes, bread crumbs	30 g*****************	_____tbsp(s) (_____g); _____cup(s) (_____g)
Cooking wine	30 mL***************	2 tbsp (30 mL)
Dietary supplements not in conventional food form	The maximum amount recommended, as appropriate, on the label for consumption per eating occasion or, in the absence of recommendations, 1 unit, e.g., tablet, capsule, packet, teaspoonful, etc.	_____tablet(s), _____capsule(s), _____packet(s), _____tsp(s) (_____g), etc.
Drink mixers (without alcohol)	Amount to make 240 mL drink (without ice).	_____fl oz (_____mL)
Chewing gum[9]	3 g*****************	_____piece(s) (_____g)
Meat, poultry and fish coating mixes, dry; seasoning mixes, dry e.g., chili seasoning mixes, pasta salad seasoning mixes	Amount to make one reference amount of final dish	_____tsps (_____g); _____tbsp(s) (_____g)
Salad and potato toppers, e.g., salad crunchies, salad crispins, substitutes for bacon bits	7 g*****************	_____tbsp(s) (_____g)
Salt, salt substitutes, seasoning salts (e.g., garlic salt)	1 g*****************	_____tsps (_____g); _____piece(s) (_____g) for discrete pieces (e.g., individually packaged products)
Spices, herbs (other than dietary supplements)	1/4 tsp or 0.5 g if not measurable by teaspoon	1/4 tsp (_____g); _____piece(s) (_____g) if not measurable by teaspoons (e.g., bay leaf)

Chart VI.E.6
(Continued)

Chapter VI

Product Category	Reference Amount	Label Statement[5]
Mixed Dishes:	**********************	
Measurable with cup, e.g., casseroles, hash, macaroni and cheese, pot pies, spaghetti with sauce, stews, etc.	1 cup***************	1 cup (____g)
Not measurable with cup, e.g., burritos, egg rolls, enchiladas, pizza, pizza rolls, quiche, all types of sandwiches.	140 g, add 55 g for products with gravy or sauce topping, e.g., enchilada with cheese sauce, crepe with white sauce[14].	____piece(s) (____g) for discrete pieces; ____ fractional slice (____g) for large discrete units
Nuts and Seeds:	**********************	
Nuts, seeds, and mixtures, all types: sliced, chopped, slivered, and whole	30 g*****************	____piece(s) (____g) for large pieces (e.g., unshelled nuts); ____tbsp(s) (____g); cup(s)(____g) for small pieces (e.g., peanuts, sunflower seeds)
Nuts and seed butters, pastes, or creams	2 tbsp**************	2 tbsp (____g)
Coconut, nut and seed flours	15 g*****************	____tbsp(s) (____g); ____cup (____g)
Potatoes and Sweet Potatoes/Yams:	**********************	
French fries, hash browns, skins or pancakes	70 g prepared; 85 g for frozen unprepared french fries	____ piece(s) (____g) for large distinct pieces (e.g., patties, skins); 2.5 oz (70 g/____ pieces) for prepared fries; 3 oz (84 g/____ pieces) for unprepared fries
Mashed, candied, stuffed, or with sauce	140 g***************	____piece(s) (____g) for discrete pieces (e.g., stuffed potato); ____cup(s) (____g)

Adopted from 21 CFR Parts 100 through 169 (4/1/06 Edition)

Product Category	Reference Amount	Label Statement[5]
Potatoes and Sweet Potatoes/Yams: *(Continued)*	*********************	
Plain, fresh, canned, or frozen	110 g for fresh or frozen; 125 g for vacuum packed; 160 g for canned liquid	____piece(s) (____g), for discrete pieces; ____cup(s) (____g) for sliced or chopped products
Salads:	*********************	
Gelatin salad	120 g**************	____cup (____g)
Pasta or potato salad	140 g**************	____cup(s) (____g)
All other salads, e.g., egg, fish, shellfish, bean, fruit, or vegetable salads	100 g**************	____cup(s) (____g)
Sauces, Dips, Gravies and Condiments:	*********************	
Barbecue sauce, hollandaise sauce, tartar sauce, other sauces for dipping (e.g., mustard sauce, sweet and sour sauce), all dips (e.g., bean dips, dairy-based dips, salsa)	2 tbsp**************	2 tbsp (____g); 2 tbsp (30 mL)
Major main entree sauces, e.g., spaghetti sauce	125 g**************	____cup (____g); ____cup (____mL)
Minor main entree sauces (e.g., pizza sauce[14], pesto sauce), other sauces used as toppings (e.g., gravy, white sauce, cheese sauce), cocktail sauce	1/4 cup************	1/4 cup (____g); 1/4 cup (60 mL)
Major condiments, e.g., catsup, steak sauce, soy sauce, vinegar, teriyaki sauce, marinades	1 tbsp**************	1 tbsp (____g); 1tbsp (15 mL))
Minor condiments, e.g., horseradish, hot sauces, mustards, Worcestershire sauce	1 tsp**************	1 tsp (____g); 1tsp (5 mL)

Chart VI.E.6
(Continued)

Chapter VI

Product Category	Reference Amount	Label Statement[5]
Snacks:	*********************	
All varieties, chips, pretzels, popcorn, extruded snacks, fruit-based snacks (e.g., fruit chips), grain based snack mixes	30 g*****************	____cup(s) (____g) for small pieces (e.g., popcorn); ____piece(s) (____g) for large pieces (e.g., large pretzels; pressed dried fruit (sheet); 1 oz (28 g/visual unit of measure) for bulk products (e.g., potato chips)
Soups:	*********************	
All varieties	245 g***************	____cup (____g); ____cup (____mL)
Sugars and Sweets:	*********************	
Baking candies (e.g., chips)	15 g*****************	____ piece(s) (____g) for large pieces: ____tbsp(s) (____g) for small pieces; 1/2 oz (14 g/visual unit of measure) for bulk products
Hard candies, breath mints	2 g*****************	____piece(s) (____g)
Hard candies, roll-type, mini-size in dispenser packages	5 g*****************	____piece(s) (____g)
Hard candies, others	15 g*****************	____piece(s) (____g) for large pieces; tbsp(s) (____g) for "mini-size" candies measurable by tablespoon; 1/2 oz (14 g/visual unit of measure) for bulk products
All other candies	40 g*****************	____piece(s) (____g); 1 1/2 oz (42 g/visual unit of measure) for bulk products
Confectioner's sugar	30 g*****************	____cup (____g)
Honey, jams, jellies, butter, molasses	1 tbsp***************	1 tbsp (____); 1 tbsp (15 mL)

Adopted from 21 CFR Parts 100 through 169 (4/1/06 Edition)

Product Category	Reference Amount	Label Statement[5]
Sugars and sweets: *(Continued)*	************************	
Marshmallows	30 g*****************	____cup(s) (____g) for small pieces: ____piece(s) (____g) for large pieces
Sugar	4 g*****************	____tsp (____g); ____piece(s) (____g) for discrete pieces (e.g., sugar cubes, individually packaged products)
Sugar substitutes	An amount equivalent to one reference amount for sugar in sweetness	____tsp(s) (____g) for solids; ____drop(s) (____g) for liquid; ____pieces(s) (____) e.g., individually packaged products
Syrups	30 mL for syrups used primarily as an ingredient (e.g., light or dark corn syrup); 60 mL for all others	2 tbsp (30 mL) for syrups used primarily as an ingredient; 1/4 cup (60 mL) for all others
Vegetables:	*********************	
Vegetables primarily used for garnish or flavor, e.g., pimento, parsley	4 g*****************	____piece(s) (____g); ____tbsp(s) (____g) for chopped products
Chili pepper, green onion	30 g*****************	____piece(s) (____g);[13] ____tbsp(s) (____g); ____cup(s) (____g) for sliced or chopped products
All other vegetables without sauce: fresh, canned, or frozen	85 g for fresh or frozen; 95 g for vacuum packed; 130 g for canned in liquid, cream-style corn, canned or stewed tomatoes, pumpkin, or winter squash	____piece(s) (____g) for large pieces (e.g., Brussels sprouts); ____cup(s) (____g) for small pieces (e.g., cut corn, green peas); 3 oz (84 g/visual unit of measure) if not measurable by cup

Chart VI.E.6	Chapter VI
(Continued)	

Product Category	Reference Amount	Label Statement[5]
Vegetables: *(Continued)*	*********************	
All other vegetables with sauce: fresh, canned or frozen	110 g***************	____piece(s) (____g) for large pieces (e.g., Brussels sprouts); ____cup(s) (____g for small pieces (e.g., cut corn, green peas); 4 oz (112 g/visual unit of measure) if not measurable by cup
Vegetable juice	240 mL***********	8 fl oz (240 mL)
Olives[11]	15 g*************	____pieces(s) (____g); ____tbsp(s) (____g) for sliced products
Pickles, all types[11]	30 g*************	1 oz (28 g/visual unit of measure)
Pickle relishes	15 g*************	____tbsp (____g)
Vegetable pastes, e.g., tomato paste	30 g*************	____tbsp (____g)
Vegetable sauces or purees, e.g., tomato sauce, tomato puree	60 g*************	____cup (____g); ____cup (____mL)

Footnotes

1. These values represent the amount (edible portion) of food customarily consumed per eating occasion and were primarily derived from the 1977-1978 and the 1987-1988 Nationwide Food Consumption Surveys conducted by the U.S. Department of Agriculture.

2. Unless otherwise noted in the Reference Amount column, the reference amounts are for the ready-to-serve or almost ready-to-serve form or the product (i.e., heat and serve, brown and serve). If not listed separately, the reference amount for the unprepared form (e.g., dry mixes; concentrates; dough; batter; dry, fresh and frozen pasta) is the amount required to make the reference amount of the prepared form. Prepared means prepared for consumption (e.g., cooked).

3. Manufacturers are required to convert the reference amount to the label serving size in a household measure most appropriate to their specific product using the procedures in *21 CFR 101.9(b)*.

4. Copies of the list of products for each product category are available from the Office of Nutritional Products, Labeling and Dietary Supplements (HFS-810), Center for Food Safety and Applied Nutrition, Food and Drug Administration, 200 C St., S.W., Washington, DC 20204.

Chart VI.E.6
(Continued)

5. The label statements are meant to provide guidance to manufacturers on the presentation of serving size information on the label, but they are not required. The term "piece" is used as a generic description of a discrete unit. Manufacturers should use the description of a unit that is most appropriate for the specific product (e.g., sandwich for sandwiches, cookie for cookies, and bar for ice cream bar). The guidance provided is for the label statement of products in ready-to-serve or almost ready-to-serve form. The guidance does not apply to the products which require further preparation for consumption (e.g., dry mixes, concentrates) unless specifically stated in the product category, reference amount, or label statement column that it is for these forms or the product. For products that require further preparation, manufacturers must determine the label statement following the rules in *21 CFR 101.9(b)* using the reference amount determined according to *21 CFR 101.12(c)*.

6. Includes cakes that weigh 10g or more per cubic inch.

7. Includes cakes that weigh 4g or more per cubic inch but less than 10g per cubic inch.

8. Includes cakes that weigh less than 4g per cubic inch.

9. Label serving size for ice cream cones and eggs of all sizes will be 1 unit. Label serving size of all chewing gums that weigh more than the reference amount that can reasonably be consumed at a single-eating occasion will be 1 unit.

10. Animal products not covered under the Federal Meat Inspection Act or the Poultry Products Inspection Act, such as flesh products from deer, bison, rabbit, quail, wild turkey, geese, ostrich, etc.

11. If packed or canned in liquid, the reference amount is for the drained solids, except for products in which both the solids and liquids are customarily consumed (e.g., canned chopped clam in juice).

12. The reference amount for the uncooked form does not apply to raw fish in *§101.45* or to single-ingredient products that consist of fish or game meat as provided for in *21 CFR 101.9(j)(11)*.

13. For raw fruit, vegetables and fish, manufacturers should follow the label statement for the serving size specified in Appendices A and B to the regulation entitled "Food Labeling; Guidelines for Voluntary Nutrition Labeling; and Identification of the 20 Most Frequently Consumed Raw Fruits, Vegetables and Fish; Definition of Substantial Compliance; Correction" (56 FR 60880, as amended 57 FR 8174, March 6, 1992).

14. Pizza sauce is part of the pizza and is not considered to be sauce topping.

Products with separately packaged ingredients or foods, consisting of an assortment of foods, or food to which other ingredients are added by the user.

h) Products with separately packaged ingredients or foods, with assortments of food, or to which other ingredients are added by the user may be labeled as follows:

(1) If a product consists of two or more separately packaged ingredients enclosed in an outer container or of assortments of the same type of food (e.g., assorted nuts or candy mixtures) in the same retail package, nutrition labeling shall be located on the outer container or retail package (as the case may be) to provide information for the consumer at the point of purchase. However, when two or more food products are simply combined together in such a manner that no outer container is used, or no outer label is available, each product shall have its own nutrition information, e.g., two boxes taped together or two cans combined in a clear plastic overwrap. When separately packaged ingredients or assortments of the same type of food are intended to be eaten at the same time, the nutrition information may be specified per serving for each component or as a composite value.

(2) If a product consists of two or more separately packaged foods that are intended to be eaten individually and that are enclosed in an outer container (e.g., variety packs of cereals or snack foods), the nutrition information shall:

(i) Be specified per serving for each food in a location that is clearly visible to the consumer at the point of purchase; and

(ii) Be presented in separate nutrition labels or in one aggregate nutrition label with separate columns for the quantitative amount by weight and the percent Daily Value for each food.

(3) If a package contains a variety of foods, or an assortment of foods, and is in a form intended to be used as a gift, the nutrition labeling shall be in the form required by paragraphs *(a)* through *(f)* of this section, but it may be modified as follows:

(i) Nutrition information may be presented on the label of the outer package or in labeling within or attached to the outer package.

(ii) In the absence of a reference amount customarily consumed in *§101.12(b)* that is appropriate for the variety or assortment of foods in a gift package, 1 ounce for solid foods, 2 fluid ounces for nonbeverage liquids (e.g., syrups) and 8 fluid ounces for beverages may be used as the standard serving size for purposes of nutrition labeling of foods subject to this paragraph. However, the reference amounts customarily consumed in *§101.12(b)* shall be used for purposes of evaluating whether individual foods in a gift package qualify for nutrient content claims or health claims.

(iii) The number of servings per container may be stated as "varied."

(iv) Nutrition information may be provided per serving for individual foods in the package, or, alternatively, as a composite per serving for reasonable categories of foods in the package having similar dietary uses and similar significant nutritional characteristics. Reasonable categories of foods may be used only if accepted by FDA. In determining whether a proposed category is reasonable, FDA will consider whether the values of the characterizing nutrients in the foods proposed to be in the category meet the compliance criteria set forth in paragraphs *(g)(3)* through *(g)(6)* of this section. Proposals for such categories may be submitted in writing to the Office of Nutritional Products, Labeling, and Dietary Supplements (HFS-800), Center for Food Safety and Applied Nutrition, Food and Drug Administration, 5100 Paint Branch Pkwy, College Park, MD 20740.

(v) If a food subject to paragraph *(j)(13)* of this section because of its small size is contained in a gift package, the food need not be included in the determination of nutrition information under paragraph *(h)* of this section if it is not specifically listed in a promotional catalogue as being present in the gift package, and:

(A) It is used in small quantities primarily to enhance the appearance of the gift package; or

(B) It is included in the gift package as a free gift or promotional item.

(4) If a food is commonly combined with other ingredients or is cooked or otherwise prepared before eating, and directions for such combination or preparations are provided, another column of figures may be used to declare nutrition information on the basis of the food as consumed in the format required in paragraph *(e)* of this section (e.g., a dry ready-to-eat cereal may be described with one set of Percent Daily Values for the cereal as sold (e.g., per ounce), and another set for the cereal and milk as suggested in the label (e.g., per ounce of cereal and ½ cup of vitamin D fortified skim milk), and a cake mix may be labeled with one set of Percent Daily Values for the dry mix (per serving) and another set for the serving of the final cake when prepared): *Provided*, That, the type and quantity of the other ingredients to be added to the product by the user and the specific method of cooking and other preparation shall be specified prominently on the label.

Chart VI.E.8
Rounding Rule Table
for Declaring Nutrients

Nutrient/ Serving	(M) or (V)*	Core Nutrient	Units	Increment Rounding **	Insignificant Amount	Other Relevant Information ***
Calories	M	X	Cal	< 5 cal - express as zero ≤ 50 cal -express to nearest 5 cal increment > 50 cal - express to nearest 10 cal increment	< 5 cal	§101.9(c)(1)
Calories from fat	M		Cal	< 5 cal - express as zero ≤ 50 cal -express to nearest 5 cal increment > 50 cal - express to nearest 10 cal increment	< 5 cal	if < 0.5g fat: "cal from fat" not required §101.9(c)(1)(ii)
Calories from saturated fat	V		Cal	< 5 cal - express as zero ≤ 50 cal -express to nearest 5 cal increment > 50 cal - express to nearest 10 cal increment	< 5 cal	§101.9(c)(1)(iii)
Total fat	M	X	g	< 0.5g - express as zero < 5g - express to nearest 0.5g increment ≥ 5g - express to nearest 1g increment	< 0.5g	§101.9(c)(2)
Saturated fat	M		g	< 0.5g - express as zero < 5g - express to nearest 0.5g increment ≥ 5g - express to nearest 1g increment	< 0.5g	§101.9(c)(2)(i)
Trans Fat	M		g	< 0.5g - express as zero < 5g - express to nearest 0.5g increment ≥ 5g - express to nearest 1g increment	< 0.5g	§101.9(c)(2)(ii)
Polyun-saturated and monoun-saturated fat	V		g	< 0.5g - express as zero < 5g - express to nearest 0.5g increment ≥ 5g - express to nearest 1g increment	< 0.5g	§101.9(c)(2)(iii) & (iv)
	M		mg	< 2mg - express as zero 2 - 5mg - express as "less than 5 mg" > 5mg - express to nearest 5mg increment	< 2mg	§101.9(c)(3)
Sodium	M	X	mg	< 5mg - express as zero 5 - 140mg - express to nearest 5mg increment > 140mg - express to nearest 10mg increment	< 5mg	§ 101.9(c)(4)

Adopted from Food Labeling: Questions and Answers Vol. II, A Guide for Restaurants and Other Retail Establishments, 8/95, Updated 11/06

Chart VI.E.8
(Continued)

Nutrient/ Serving	(M) or (V)*	Core Nutrient	Units	Increment Rounding **	Insignificant Amount	Other Relevant Information ***
Potassium	V		mg	< 5mg - express as zero 5 - 140mg - express to nearest 5mg increment > 140mg - express to nearest 10mg increment	< 5mg	§ 101.9(c)(5)
Total carbo-hydrate	M	X	g	< 0.5g - express as zero < 1g - express as "Contains less than 1g" OR "less than 1g" = 1g - express to nearest 1g increment	< 1g	§ 101.9(c)(6)
Dietary fiber	M		g	< 0.5g - express as zero < 1g - express as "Contains less than 1g" OR "less than 1g" = 1g - express to nearest 1g increment	< 1g	§101.9(c)(6)i
Soluble & insoluble fiber	V		g	< 0.5g - express as zero < 1g - express as "Contains less than 1g" OR "less than 1g" = 1g - express to nearest 1g increment	<0.5g	§101.9(c)(6)(i) (A) & (B)
Sugars	M		g	< 0.5g - express as zero < 1g - express as "Contains less than 1g" OR "less than 1g" = 1g - express to nearest 1g increment	<0.5g	§101. 9(c)(6)(ii)
Sugar alcohol	V		g	< 0.5g - express as zero < 1g - express as "Contains less than 1g" OR "less than 1g" =1g - express to nearest 1g increment	< .5g	§101.9(c)(6)(iii)
Other carbo-hydrates	V		g	< 0.5g - express as zero < 1g - express as "Contains less than 1g" OR "less than 1g" =1g - express to nearest 1g increment	< .5g	§101.9(c)(6)(iv)
Protein	M	X	g	< 0.5g - express as zero < 1g - express as "Contains less than 1g" OR "less than 1g" =1g - express to nearest 1g increment	< 1g	§101.9(c)(7)

Adopted from Food Labeling: Questions and Answers Vol. II, A Guide for Restaurants and Other Retail Establishments, 8/95

Chart VI.E.8
(Continued)

Chapter VI

Nutrient/ Serving	(M) or (V)*	Core Nutrient	Units	Increment Rounding **	Insignificant Amount	Other Relevant Information ***
Vitamins & Minerals	M		%DV	< 2% of RDI - may be expressed as: (1) 2% if actual amount is 1.0% or more (2) zero (3) an asterisk that refers to statement "Contains less than 2% of the Daily Value of this (these) nutrient (nutrients) (4) for Vitamins A and C, calcium, iron: statement "Not a significant source of ____(listing the vitamins or minerals omitted)" = 10% or RDI - express to nearest 2% increment >10% - = 50% of RDI - express to nearest 5% increment > 50% of RDI - express to nearest 10% increment	< 2% RDI	Vitamins and minerals other than Vitamins A and C, calcium and iron, listed in *§101.9(c)(8)(iv)*, are mandatory if added as nutrient supplement in food or if claim is made *§101.9 (c)(8)(iii) & (iv)*
Beta-carotene	V		% Vit A	= 10% of RDI - express to nearest 2% increment >10% - = 50% of RDI - express to nearest 5% increment > 50% of RDI - express to nearest 10% increment		*§101.9(c)(8)(vi)*

*   (M) = Mandatory and (V) = Voluntary

**   To express to the 1g increment, amounts exactly halfway between two whole numbers or higher (e.g., 2.50 to 2.66g) round up (e.g., 3g) and amounts less than halfway between two whole numbers (e.g., 2.01 to 2.49g) round down (e.g., 2g).

***   NOTES FOR ROUNDING % Daily Value (DV)
(1)   To calculate %DV, divide either the actual (unrounded) quantitative amount or the declared (rounded) amount by the appropriate RDI or DRV. Use whichever amount will provide the greatest consistency on the food label and prevent unnecessary consumer confusion. *(21 CFR 101.9(d)(7)(ii))*.

(2)   When %DV values fall between two whole numbers, rounding shall be as follows:
for values exactly halfway between two whole numbers and higher (e.g., 2.5 to 2.99) round up (e.g., 3%).
For values less than halfway between two whole numbers (e.g., 2.01 to 2.49) round down (e.g., 2%).

Adopted from Food Labeling: Questions and Answers Vol. II, A Guide for Restaurants and Other Retail Establishments, 8/95

Chart VI.E.9
Rounding Rule Table
for Serving Size

Measurement Type	Unit	Increment Rounding
Discrete Units	Servings	≤ 50% of the reference amounts(RA): number of units closest to the RA = 1 serving
		= 50% to < 67% of RA: then 1 unit - 1 serving **OR** 2 units = 1 serving
		≥ 67% to < 200% of RA: then 1 unit = 1 serving
		≥ 200% of RA: then 1 unit = 1 serving, if it can reasonably be consumed at a single eating occasion
Common Household Measures	Volume: Cup (cup)	Use "cup" in $1/3$ or $1/4$ cup increments, except may use "fl oz" for beverages
	Tablespoon (tbsp)	≥ 2 tbsp & < ¼ cup = whole tbsp
		Between 1 & 2 tbsp, may use increments of: 1, $1^1/3$, $1^1/2$, $1^2/3$, 2
	Teaspoon (tsp)	≥ 1 tsp & < 1 tbsp = whole tsp < 1 tsp = $1/4$ tsp increments
	Fluid ounce (fl oz)	Fluid ounce (fl oz) = whole number increments
	Weight: Ounce (oz)	Ounce (oz) measures = 0.5oz increments
		Serving sizes that fall half-way between two serving sizes, manufacturers shall round up to the next incremental size

Metric Measures	Volume:
	    Milliliters (mL)
	Weight:
	    Gram (g)
	    Milligram (mg)
		> 5 = nearest whole number
		≥ 2 and < 5 = nearest 0.5
		
		< 2 = nearest 0.1

Adopted from Food Labeling: Questions & Answers, Foods Other than Dietary Supplements, FDA, 8/93

Chart VI.E.9
(Continued)

Chapter VI

Measurement Type	Unit	Increment Rounding
Number of Servings/Container	Numbers	Round to the nearest whole number except for servings between 2 and 5 servings Between 2 and 5 servings = nearest 0.5 serving Rounding should be indicated by the term "about"

Adopted from Food Labeling: Questions & Answers, Foods Other than Dietary Supplements, FDA, 8/93

Chart VI.E.10
Nutrition Labeling
Summary Sheet

Nutrient	(M) or (V)[1]	Printing		If Mandatory, When May Be Deleted	If Voluntary, When Becomes Mandatory	Simple Definition
		Bold[1]	Indent[1]			
Calories (Total Calories or Calories Total) §101.9(c)(1)	M	Y	N	Never		Calculated, See §101.9(c)(1)(i)
Calories from Fat §101.9(c)(1)(ii)	M	N	Y (In standard vertical display should follow "Calories" on same line)	If less than 0.5g fat/serving and statement "Not a significant source of Calories from Fat" is at bottom of nutrient list.		
Calories from Sat Fat §101.9(c)(1)(iii)	V	N	Y		Never	
Total Fat (Fat Total) §101.9(c)(2)	M	Y	N	Never		Total lipid fatty acids expressed as triglycerides.
Saturated Fat (Saturated) §101.9(c)(2)(i)	M	N	Y	If less than 0.5g Total Fat/Serving and no claim is made about fat, fatty acid, or cholesterol, and if "Calories from Sat Fat" is not declared. If deleted, statement "Not a significant source of saturated fat" must follow nutrient list.		Sum of all fatty acids containing no double bonds.
Trans Fat §101.9(c)(2)(ii)	M	N	Y	If less than 0.5g Total Fat/Serving and no claim is made about fat, fatty acid, or cholesterol. If deleted, statement "Not a significant source of trans fat" must follow nutrient list.		Sum of all unsaturated fatty acids that contains one or more isolated (i.e., nonconjugated) double bonds in a trans configuration.
Polyunsaturated Fat (Polyunsaturated) §101.9(c)(2)(iii)	V	N	Y		When monounsaturated fat is declared, or when claim about fatty acids or cholesterol is made unless food is "fat free".	cis - cis methylene interrupted polyunsaturated fatty acid.

Adopted from Food Labeling: Questions and Answers Vol. II, A Guide for Restaurants and Other Retail Establishments, 8/95, Updated 11/06

107

Chart VI.E.10
(Continued)

Chapter VI

Nutrient	(M) or (V)[1]	Printing Bold[1]	Printing Indent[1]	If Mandatory, When May Be Deleted	If Voluntary, When Becomes Mandatory	Simple Definition
Monounsaturated Fat (Monounsaturated) §101.9(c)(2)(iv)	V	N	Y	—	When polyunsaturated fat is declared, or when claim about fatty acids or cholesterol is made unless food is "fat free".	*cis-*monounsaturated fatty acids.
Cholesterol §101.9(c)(3)	M	Y	N	If less than 2mg cholesterol/ serving and no claim is made about fat, fatty acids, or cholesterol. If deleted, statement "Not a significant source of cholesterol" must follow nutrient list.	---	---
Sodium §101.9(c)(4)	M	Y	N	Never	---	---
Potassium §101.9(c)(5)	V	Y	N	---	When claim is made about potassium.	---
Carbohydrate Total (Total Carbohydrate) §101.9(c)(6)	M	Y	N	Never	---	Difference between the sum of crude protein, total fat, moisture and ash and the total weight of food.
Dietary Fiber §101.9(c)(6)(i)	M	N	Y	If serving contains less than 1g (or may state less than 1g). If deleted, statement "Not a significant source of dietary fiber" must follow nutrient list.	---	---
Soluble Fiber §101.9(c)(6)(i)(A)	V	N	Y	---	When claim is made about soluble fiber.	---
Insoluble Fiber §101.9(c)(6)(i)(B)	V	N	Y	---	When claim is made about insoluble fiber.	---
Sugars §101.9(c)(6)(ii)	M	N	Y	If serving contains less than 1g of sugars and no	---	Sum of all free mono- and disaccharides

Adopted from FDA Guide to Nutritional Labeling and Education Act (NLEA) Requirements, August 1994, Updated 11/06

Chart VI.E.10
(Continued)

Nutrient	(M) or (V)[1]	Printing		If Mandatory, When May Be Deleted	If Voluntary, When Becomes Mandatory	Simple Definition
		Bold[1]	Indent[1]			
				claim is made about sweeteners, sugars, or sugar alcohol content. If deleted, statement "Not a significant source of sugars" must follow nutrient list.		(e.g., glucose, fructose, lactose, and sucrose).
Sugar Alcohol *(May state specific sugar alcohol if only one is present in the food.)* §101.9(c)(6)(iii)	V	N	Y	---	When claim is made about sugar alcohols or sugars when sugar alcohols are present in the food.	Sum of saccharide derivatives in which a hydroxyl group replaces a ketone or aldehyde group and whose use in the food is listed by FDA (e.g., mannitol or xylitol) or is GRAS (e.g., sorbitol).
Other Carbohydrate §101.9(c)(6)(iv)	V	N	Y			Difference between total carbohydrates and sum of dietary fiber, sugars and sugar alcohol. Except if sugar alcohol not <u>declared</u>, it is difference between total carbohydrates and sum of dietary fiber and sugars.
Protein §101.9(c)(7)	M	Y	N	Grams of protein never optional, always required. % DV is optional unless (1) protein claim is made, (2) food is for infants or children less than 4 years, or (3) food is for adults and children 4 or more years of age and has a protein quality (protein digestibility-corrected amino acid score) of less than 20%.		May be calculated as 6.25 times nitrogen content. See §101.9(c)(7)(ii) for method of determining protein quality (%DV).

Chart VI.E.10
(Continued)

Chapter VI

Nutrient	(M) or (V)[1]	Printing		If Mandatory, When May Be Deleted	If Voluntary, When Becomes Mandatory	Simple Definition
		Bold[1]	Indent[1]			
Vitamins and minerals				If serving contains less than 2% DV, may (1) be declared as zero, (2) be declared as 2% if that is the nearest 2% increment, (3) be deleted and replaced with "Not a significant source of ___" statement, or (4) use asterisk referring to statement "Contains less than 2% of the Daily Value of this (these) nutrient(s)".	Any other vitamin or mineral with an RDA that is added as a nutrient supplement or when claim is made about them. Not required on label if required or permitted by food standard and that standardized food is included as an ingredient, or if nutrient is included solely for technical purposes and is declared only in the ingredient statement .	Expressed as % DV
Vitamin A	M	N	N			
Vitamin C	M	N	N			
Calcium	M	N	N			
Iron	M	N	N			
Any other vitamin or mineral listed in *§101.9(c)(8)(iv)* (in order given)	V	N	N (can be listed horizontally or in columns)			
% of Vitamin A as beta-carotene *§101.9(c)(8)(iv)*	V	N	Y (or in parenthesis)			

1. M = mandatory   V = voluntary   Y = yes   N = no

Adopted from FDA Guide to Nutritional Labeling and Education Act (NLEA) Requirements, August 1994, Updated 11/06

The purpose of these guidelines is to assist manufacturers in determining the metric equivalent declarations (e.g., gram (g) and milliliter (mL) measures) of the common household measures that are declared on food labels. When FDA performs nutrient analyses to determine the accuracy of nutrition labeling, assessment of compliance is based on these metric quantities.

The Nutrition Labeling and Education Act of 1990 added section 403(q) to the Federal Food, Drug, and Cosmetic Act (21 U.S.C. 343 (q). This section specifies, in part, that the serving size is "an amount customarily consumed ... expressed in a common household measure that is appropriate to the food," or "if the use of the food is not typically expressed in a serving size, the common household unit of measure that expresses the serving size of the food" should be used. For example, for a product such as pancake mix that is an ingredient of a food, if $1/4$ cup of pancake mix is required to make the customarily consumed amount of pancakes, the serving size of this pancake mix would be expressed as $1/4$ cup of mix.

Serving sizes are determined from the reference amounts established in *21 CFR 101.12(b)* and the procedures described in *21 CFR 101.9(b)(2)* and must be expressed in both common household measures and equivalent metric quantities (*21 CFR 101.9(b)(7)*). As stated in *21 CFR 101.9(b)(5)* the term "common household measure" or "common household unit" means cup, tablespoon (tbsp), teaspoon (tsp), piece, slice, fraction (e.g., $1/4$ pizza), ounce (oz), fluid ounce (fl oz), or other common household equipment used to package food products (e.g., jar, tray).

For specific details of the final rules that apply to serving sizes, refer to the following sections of the Code of Federal Regulations (CFR):

21 CFR 101.9(b)	Nutrition labeling of food; definition of serving sizes
21 CFR 101.9(b)(6)	Single-serving containers
21 CFR 101.9(b)(8)	Number of servings per container
21 CFR 101.12	Reference amounts customarily consumed per eating occasion

General Information

1. Representative samples of a food should be selected using standard sampling techniques from various lots (Ref. *21 CFR 101.9(g)(2)*). For mixtures (e.g., solids in solids, such as brownies with nuts; solids in liquids, such as soup with vegetables, the sample selected should contain a representative amount of the incorporated solids.

2. Good quality laboratory equipment (e.g., graduated cylinders, balances, etc.) should be used to measure or weigh the food. Equipment should be calibrated in accordance with good laboratory practices and/or manufacturer's specifications.

3. Standard analytical practices should be used for accurately determining product weights and volumes. Significant digits should be retained in order to minimize rounding errors in reporting final values.

4. Each set of measurements should be determined by the same trained operator using the same methodology (i.e., the same equipment, procedures, and techniques under the same conditions. For variable products (e.g., small pastas, snacks), another set of measurements should be determined by a second individual.

5. All measurements should be replicated a sufficient number of times to ensure that the average of the measurements is representative of the product.

6. Foods and containers should be at appropriate and compatible temperatures for volume determination. Foods stored at room temperature should be measured at 20° C, refrigerated foods should be at 4° C, and frozen foods should be measured at the frozen temperature.

7. The quality of the food product should be maintained throughout. Moisture gains or losses should be minimized. Fragile products should be handled carefully to minimize product breakdown. For example, flake breakfast cereals should be carefully transferred to volumetric containers and should not be sifted, stirred, or packed. Measurements should be made prior to excessive handling or shipping.

8. The food volume measured should be at least 10 times the reference amount for the category in order to minimize measuring errors. (For example, dividing the weight of a cup of a product by 16 and 48 provides the tablespoon and teaspoon weights, respectively.)

9. For purposes of nutrition labeling, 1 cup means 240 mL, 1 tablespoon means 15 mL, 1 teaspoon means 5 mL, 1 fluid ounce means 30 mL, and 1 ounce means 28 g *(21 CFR 101.9(b)(5)(viii))*.

10. As defined in *21 CFR 101.9(b)(5)(i)*, the household measures of cups, tablespoons, or teaspoons should be used whenever possible. Fluid ounces may be used for beverages. These measures should be expressed as follows:

Measure	**Increment**
Cups	$^1/_4$ or $^1/_3$-cup increments.
Tablespoons	(1) Use whole numbers of tablespoons for quantities $< ^1/_4$ cup but ≥ 2 tablespoons.
	(2) Use 1, $1^1/_3$, $1^1/_2$, $1^2/_3$ tablespoons for quantities < 2 tablespoons but $\geq$ 1 tablespoon.
Teaspoons	(1) Use whole numbers of teaspoons for quantities < 2 tablespoons but $\geq$ 1 teaspoon.
	(2) Use $^1/_4$ teaspoon increments for quantities < 1 teaspoon.

    If cups, tablespoons, or teaspoons, are not applicable, units such as piece, slice, tray, jar, and fractions should be used *(21 CFR I01.9(b)(5)(ii))*. The fractional slice of a food that most closely approximates the reference amount should be expressed as follows *(21 CFR 101.9(b)(2)(ii))*:

Fractions:	$^1/_2$, $^1/_3$, $^1/_4$, $^1/_5$, or $^1/_6$, and smaller fractions that can be generated by further divisions by 2 or 3 (such as $^1/_8$, $^1/_9$, $^1/_{10}$, $^1/_{12}$, $^1/_{15}$, $^1/_{16}$, etc.).

Adopted from letter (10/1/93) – Office of Food Labeling, Center for Food Safety and Applied Nutrition

If other units are not applicable, ounce may be used and must be accompanied by an appropriate visual unit of measure, for example, 1 oz (28 g, about 1 inch slice of cheese) *(21 CFR 101.9(b)(5)(iii))*.

11. When the serving size is exactly half way between two values, it should be rounded to the higher value *(21 CFR 101.9(b)(5)(ix))*, for example, 2.5 tbsp is rounded to 3 tbsp.

12. Grams and milliliters should be rounded to the nearest whole number except for quantities that are less than 5 g or 5 mL. Gram and milliliter quantities between 2 and 5 should be rounded to the nearest 0.5 g or mL. Gram and milliliter quantities less than 2 should be expressed in 0.1 g (mL) increments *(21 CFR 101.9(b)(7)(ii))*.

13. The provision in *21 CFR 101.9(b)(7)* exempts single-serving containers from listing metric equivalent except when nutrition information is on a drained weight basis in accordance with *21 CFR 101.9(h)(9)*. If companies voluntarily list metric equivalents for single-serving containers, the value must agree with the net quantity of contents expression.

14. FDA is unaware of any need to make changes in the procedures for determining metric equivalents of household measures due to the effects of variations in altitude. The agency will consider the need for altitude corrections should data become available.

Liquid Measurements

1. Liquids may be measured in volumetric glassware or graduated cylinders. The level should be read at the lowest part of the meniscus and care should be taken to avoid parallax error. For clear liquids, a shade or dark material behind the meniscus may improve observation.

2. The volume being measured should be within 25 percent of the total capacity of the glassware selected. Select the smallest container that will hold the intended volume.

3. Techniques for determining the volumes of viscous liquids (e.g., syrups, molasses), fluid type solids (e.g., applesauce, hot breakfast cereals), and spoonable thick or gelatinous solid-type liquids (e.g., gelatins, mayonnaise) include direct fill and volume displacement:

    a. Direct fill—Direct fill involves carefully transferring the product to avoid incorporating air bubbles, allowing time for settling (viscous liquids are higher in the center when first poured), and, if necessary, leveling with a straight edge or by extrusion. For example, for a hot breakfast cereal product, a density cup of known weight and volume may be used to determine the volume to weight relationship: the cereal is transferred to the measuring container, the sliding disk is moved into position leaving a small gap, the excess cereal escapes through the opening and is wiped away, and the weight of the known volume of cereal can be determined by difference.

    b. Volume displacement--Volume displacement involves adding a second material to fill the air space above the product. For example, a measured amount of water can be added to completely fill the air space above the mayonnaise in a mayonnaise jar. The volume of mayonnaise can be determined as the difference between the volume of the jar and the volume of water added. The material selected should not mix with the product being measured.

Solid Measurements

1. Fine particulate solids (for example, sugars, batter mixes, flours) may be leveled using a knife or other straight-edge after transfer to an appropriate volumetric measure (e.g., a cup measure would be appropriate for determination of tablespoons or teaspoons).

2. Medium particulate solids (e.g., nuts, flakes, pastas) should have the particle volume above the fill line approximately equal to the free air space found between particles immediately below the fill line.

3. For products where the packing liquid is not normally consumed (for example, olives, pickles, tuna fish, etc.), products should be drained for 2 minutes on a No. 8 sieve before weighing or measuring. AOAC procedures for canned vegetables and fish products are described in sections 968.30 and 937.07, respectively, of *Official Methods of Analysis* (Reference 1).

4. Products should be measured in the form in which they are packaged and sold (see 6 above under General Information). Some frozen products (e.g., frozen blocks of vegetables, frozen juice concentrates) cannot be transferred to volumetric containers in the frozen state. This type of product may need to be broken apart and/or defrosted slightly in order to fit into the measuring container. If necessary, cover and thaw the product minimally, transfer to measuring container, and return the product to its frozen state for measurement.

5. Techniques for determining the volumes of bulk solids (e.g., bulk cheeses and irregularly-shaped solids (e.g., ice cream novelties) include direct measurement and volume displacement:

   a. Direct measurement — Direct measurement involves creating a representative piece with regular dimensions. For example, the dimensions of a one-ounce cube of cheese can be directly measured with a ruler.

   b. Volume displacement — Volume displacement involves immersing the irregularly-shaped object in a known volume of another material and measuring the amount of material displaced. For example, an ice cream bar can be dipped briefly into a vat of cold liquid. The volume of the displaced liquid can be determined directly or by difference. The amount of displaced liquid is a measure of the volume of the irregularly-shaped ice cream bar.

Reference
1. *Official Methods of Analysis*, 15th Ed., AOAC, Arlington, VA (1990).

Adopted from letter (10/1/93) – Office of Food Labeling, Center for Food Safety and Applied Nutrition

Chart VI.E.12
Daily Values for Nutrition Labeling

(Based on 2,000 Calorie Intake for Adults and Children 4 or More Years of Age)

Nutrients in this table are listed in the order in which they are required to appear on a label in accordance with *(21 CFR 101.9(c))*.

This list includes only those nutrients for which a Daily Reference Value (DRV) has been established in *§ 101.9(c)(9)* or a Reference Daily Intake (RDI) in *§ 101.9(c)(8)(iv)*.

M = Mandatory
V = Voluntary

* Grams of protein never optional, always required. %DV is optional unless (1) protein claim is made, (2) food is for infants or children less than 4 years, or (3) food is for adults and children 4 or more years of age and has a protein quality (protein digestibility corrected amino acid score) of less than 20%.

NUTRIENT	M OR V	UNIT OF MEASURE	DAILY VALUE
Total Fat	M	Grams (g)	65
Saturated Fatty Acids	M	Grams (g)	20
Cholesterol	M	Milligrams (mg)	300
Sodium	M	Milligrams (mg)	2,400
Potassium	V	Milligrams (mg)	3,500
Total carbohydrate	M	Grams (g)	300
Dietary Fiber	M	Grams (g)	25
Protein	M*	Grams (g)	50
Vitamin A	M	International Units (IU)	5,000
Vitamin C	M	Milligrams (mg)	60
Calcium	M	Milligrams (mg)	1,000
Iron	M	Milligrams (mg)	18
Vitamin D	V	International Units (IU)	400
Vitamin E	V	International Units (IU)	30
Vitamin K	V	Micrograms (μg)	80
Thiamin	V	Milligrams (mg)	1.5
Riboflavin	V	Milligrams (mg)	1.7
Niacin	V	Milligrams (mg)	20
Vitamin B_6	V	Milligrams (mg)	2.0
Folate	V	Micrograms (μg)	400
Vitamin B_{12}	V	Micrograms (μg)	6.0
Biotin	V	Micrograms (μg)	300
Pantothenic acid	V	Milligrams (mg)	10
Phosphorus	V	Milligrams (mg)	1,000
Iodine	V	Micrograms (μg)	150
Magnesium	V	Milligrams (mg)	400
Zinc	V	Milligrams (mg)	15
Selenium	V	Micrograms (μg)	70
Copper	V	Milligrams (mg)	2.0
Manganese	V	Milligrams (mg)	2.0
Chromium	V	Micrograms (μg)	120
Molybdenum	V	Micrograms (μg)	75
Chloride	V	Milligrams (mg)	3,400

Illustration VI.E.13
Illustration of Type and Point Sizes

Chapter VI

Helvetica Regular 8 point with 1 point of leading

Franklin Gothic Heavy or Helvetica Black, flush left & flush right, no smaller than 13 point

3 point rule

8 point Helvetica Black with 4 points of leading

7 point rule

¼ point rule centered between nutrients (2 points leading above and 2 points below)

6 point Helvetica Black

All labels enclosed by ½ point box rule within 3 points of text measure

8 point Helvetica Regular with 4 points of leading

¼ point rule

8 point Helvetica Regular, 4 points of leading with 10 point bullets

Type below vitamins and minerals (footnotes) is 6 point with 1 point of leading

Nutrition Facts

Serving Size 1 cup (228g)
Servings Per Container 2

Amount Per serving

Calories 260   Calories from Fat 120

% Daily Value*

Total Fat 13 g	**20%**
Saturated Fat 5g	**25%**
Trans Fat 2g	
Cholesterol 30mg	**10%**
Sodium 660mg	**28%**
Total Carbohydrate 31g	**11%**
Dietary Fiber 0g	**0%**
Sugars 5g	
Protein 5g	

Vitamin A 4%	•	Vitamin C 2%	
Calcium 15%	•	Iron 4%	

* Percent Daily Values are based on a 2,000 calorie diet. Your daily values may be higher or lower depending on your calorie needs:

		Calories:	2,000	2,500
Total Fat	Less than		65g	80g
Sat Fat	Less than		20g	25g
Cholesterol	Less than		300mg	300mg
Sodium	Less than		2,400mg	2,400mg
Total Carbohydrate			300g	375g
Dietary Fiber			25g	30g

Adopted from 21 CFR Parts 100 through 169 (4/1/06 Edition)

A. Overall

1. The Nutrition Facts label is boxed with all black or one color type printed on a white or neutral ground.

B. Typeface and Size

1. The Nutrition Facts label uses 6 point or larger Helvetica Black and/or Helvetica Regular type. In order to fit some formats the typography may be kerned as much as -4, (tighter kerning reduces legibility).

2. Key nutrients & their % Daily Value are set in 8 point Helvetica Black (but "% is set in Helvetica Regular).

3. "Nutrition Facts" is set in either Franklin Gothic Heavy or Helvetica Black to fit the width of the label flush left and flush right.

4. "Serving Size" and "Servings per container" are set in 8 point Helvetica Regular with 1 point of leading.

5. The table labels (for example; "Amount per Serving") are set 6 point Helvetica Black.

6. Absolute measures of nutrient content (for example; "1g") and nutrient subgroups are set in 8 point Helvetica Regular with 4 points of leading.

7. Vitamins and minerals are set in 8 point Helvetica Regular, with 4 points of leading, separated by 10 point bullets.

8. All types that appear under vitamins and minerals are set in 6 point Helvetica regular with 1 point of leading.

C. Rules

1. A 7 point rule separates large groupings as shown in example. (See Chart VI.E.13(i)) A 3 point rule separates calorie information from the nutrient information.

2. A hairline rule or 1/4 point rule separates individual nutrients, as shown in the example. Descenders do not touch rule. The top half of the label (nutrient information) has 2 points of leading between the type and the rules, the bottom half of the label (footnotes) has 1 point of leading between the type and the rules.

D. Box

1. All labels are enclosed by 1/2 point box rule within 3 points of text measure.

j) The following foods are exempt from this section or are subject to special labeling requirements:
  (1) (i) Food offered for sale by a person who makes direct sales to consumers (e.g., a retailer) who has annual gross sales made or business done in sales to consumers that is not more than $500,000 or has annual gross sales made or business done in sales of food to consumers of not more than $50,000, *Provided,* That the food bears no nutrition claims or other nutrition information in any context on the label or in labeling or advertising. Claims or other nutrition information subject the food to the provisions of this section.
    (ii) For purposes of this paragraph, calculation of the amount of sales shall be based on the most recent 2-year average of business activity. Where firms have been in business less than 2 years, reasonable estimates must indicate that annual sales will not exceed the amounts specified. For foreign firms that ship foods into the United States, the business activities to be included shall be the total amount of food sales, as well as other sales to consumers, by the firm in the United States.
  (2) Food products which are:
    (i) Served in restaurants, *Provided,* That the food bears no nutrition claims or other nutrition information in any context on the label or in labeling or advertising. Claims or other nutrition information subject the food to the provisions of this section;
    (ii) Served in other establishments in which food is served for immediate human consumption (e.g., institutional food service establishments, such as schools, hospitals, and cafeterias; transportation carriers, such as trains and airplanes; bakeries, delicatessens and retail confectionery stores where there are facilities for immediate consumption on the premises; food service vendors, such as lunch wagons, ice cream shops, mall cookie counters, vending machines, and sidewalk carts where foods are generally consumed immediately where purchased or while the consumer is walking away, including similar foods sold from convenience stores; and food delivery systems or establishments where ready-to-eat foods are delivered to homes or offices), *Provided,* That the food bears no nutrition claims or other nutrition information in any context on the label or in labeling or advertising. Claims or other nutrition information subject the food to the provisions of this section;
    (iii) Sold only in such facilities, *Provided,* That the food bears no nutrition claims or other nutrition information in any context on the label or in labeling or advertising. Claims or other nutrition information subject the food to the provisions of this section;
    (iv) Used only in such facilities and not served to the consumer in the package in which they are received (e.g., foods that are not packaged in individual serving containers); or
    (v) Sold by a distributor who principally sells food to such facilities: *Provided,* That:
      (A) This exemption shall not be available for those foods that are manufactured, processed, or repackaged by that distributor for sale to any persons other than restaurants or other establishments that serve food for immediate human consumption, and
      (B) The manufacturer of such products is responsible for providing the nutrition information on the products if there is a reasonable possibility that the product will be purchased directly by consumers.
  (3) Food products that are:
    (i) Of the type of food described in paragraphs (j)(2)(i) and (j)(2)(ii) of this section,
    (ii) Ready for human consumption,
    (iii) Offered for sale to consumers but not for immediate human consumption,
    (iv) Processed and prepared primarily in a retail establishment, and
    (v) Not offered for sale outside of that establishment, (e.g., ready-to-eat foods that are processed and prepared on-site and sold by independent delicatessens, bakeries, or retail confectionery stores where there are no facilities for immediate human consumption; by in-store delicatessen, bakery, or candy departments; or at self-service food bars such as salad bars), *Provided,* That the food bears no nutrition claims or other nutrition information in any context on the label or in labeling or advertising. Claims or other nutrition information subject the food to the provisions of this section.

(4) Foods that contain insignificant amounts of all of the nutrients and food components required to be included in the declaration of nutrition information under paragraph (c) of this section, *Provided,* That the food bears no nutrition claims or other nutrition information in any context on the label or in labeling or advertising. Claims or other nutrition information subject the food to the provisions of this section. An insignificant amount of a nutrient or food component shall be that amount that allows a declaration of zero in nutrition labeling, except that for total carbohydrate, dietary fiber, and protein, it shall be an amount that allows a declaration of "*less than 1 gram.*" Foods that are exempt under this paragraph include coffee beans (whole or ground), tea leaves, plain unsweetened instant coffee and tea, condiment-type dehydrated vegetables, flavor extracts, and food colors.

(5) (i) Foods, other than infant formula, represented or purported to be specifically for infants and children less than 2 years of age shall bear nutrition labeling, except as provided in paragraph (j)(5)(ii) and except that such labeling shall not include calories from fat (paragraph (c)(l)(ii) of this section), calories from saturated fat ((c)(l)(iii)), saturated fat ((c)(2)(i)), polyunsaturated fat ((c)(2)(ii)), monounsaturated fat ((c)(2)(iii)), and cholesterol ((c)(3)).

   (ii) Foods, other than infant formula, represented or purported to be specifically for infants and children less than 4 years of age shall bear nutrition labeling, except that:
   (A) Such labeling shall not include declarations of percent of Daily Value or total fat, saturated fat, cholesterol, sodium, potassium, total carbohydrate, and dietary fiber;
   (B) Nutrient names and quantitative amounts by weight shall be presented in two separate columns;
   (C) The heading "*Percent Daily Value*" required in paragraph (d)(6) of this section shall be placed immediately below the quantitative information by weight for protein;
   (D) Percent of Daily Value for protein, vitamins, and minerals shall be listed immediately below the heading "*Percent Daily Value*"; and
   (E) Such labeling shall not include the footnote specified in paragraph (d)(9) of this section.

(6) Dietary supplements, except that such foods shall be labeled in compliance with §101.36, except that dietary supplements of vitamins or minerals in food in conventional form (e.g., breakfast cereals), of herbs, and of other similar nutritional substances shall conform to the labeling of this section.

(7) Infant formula subject to section 412 of the act, as amended, except that such foods shall be labeled in compliance with part 107 of this chapter.

(8) Medical foods as defined in section 5(b) of the Orphan Drug Act (21 U.S.C. 360ee(b)(3)). A medical food is a food which is formulated to be consumed or administered internally under the supervision of a physician and which is intended for the specific dietary management of a disease or condition for which distinctive nutritional requirements, based on recognized scientific principles, are established by medical evaluation. A food is subject to this exemption only if:
   (i) It is a specially formulated and processed product (as opposed to a naturally occurring foodstuff used in its natural state) for the partial or exclusive feeding of a patient by means of oral intake or internal feeding by tube;
   (ii) It is intended for the dietary management of a patient who, because of therapeutic or chronic medical needs, has limited or impaired capacity to ingest, digest, absorb, or metabolize ordinary foodstuffs or certain nutrients, or who has other special medically determined nutrient requirements, the dietary management of which cannot be achieved by the modification of the normal diet alone;
   (iii) It provides nutritional support specifically modified for the management of the unique nutrient needs that result from the specific disease or condition, as determined by medical evaluation;
   (iv) It is intended to be used under medical supervision; and
   (v) It is intended only for a patient receiving active and ongoing medical supervision wherein the patient requires medical care on a recurring basis for, among other things, instructions on the use of the medical food.

(9) Food products shipped in bulk form that are not for distribution to consumers in such form and that are for use solely in the manufacture of other foods or that are to be processed, labeled, or repacked at a site other than where originally processed or packed.

(10) Raw fruits, vegetables, and fish subject to section 403(q)(4) of the act, except that the labeling of such foods should adhere to guidelines in §101.45. This exemption is contingent on the food bearing no nutrition claims or other nutrition information in any context on the label or in labeling

or advertising. Claims or other nutrition information subject the food to nutrition labeling in accordance with §101.45. The term "*fish*" includes fresh water or marine fin fish, crustaceans, and mollusks, including shellfish, amphibians, and other forms of aquatic animal life.

(11) Packaged single-ingredient products that consist of fish or game meat (i.e., animal products not covered under the Federal Meat Inspection Act or the Poultry Products Inspection Act, such as flesh products from deer, bison, rabbit, quail, wild turkey, or ostrich) subject to this section may provide required nutrition information for a 3-ounce cooked edible portion (i.e., on an "as prepared" basis), except that:

(i) Such products that make claims that are based on values as packaged must provide nutrition information on an as packaged basis, and

(ii) Nutrition information is not required for custom processed fish or game meats.

(12) Game meats (i.e., animal products not covered under the Federal Meat Inspection Act or the Poultry Products Inspection Act, such as flesh products from deer, bison, rabbit, quail, wild turkey, or ostrich) may provide required nutrition information on labeling in accordance with the provisions of paragraph (a)(2) of this section.

(13)(i) Foods in small packages that have a total surface area available to bear labeling of less than 12 square inches, *Provided,* That the labels for these foods bear no nutrition claims or other nutrition information in any context on the label or in labeling or advertising. Claims or other nutrition information subject the food to the provisions of this section.

(A) The manufacturer, packer, or distributor shall provide on the label of packages that qualify for and use this exemption an address or telephone number that a consumer can use to obtain the required nutrition information (e.g., "*For nutrition information, call 1-800 123-4567*").

(B) When such products bear nutrition labeling either voluntarily or because nutrition claims or other nutrition information is provided, all required information shall be in type size no smaller than 6 point or all upper-case type of 1/16 inch minimum height, except that individual serving-size packages of food served with meals in restaurants, institutions, and on board passenger carriers, and not intended for sale at retail, may comply with §101.2(c)(5).

(ii) Foods in packages that have a total surface area available to bear labeling of 40 or less square inches may modify the requirements of paragraphs (c) through (f) and (i) of this section by one or more of the following means:

(A) Presenting the required nutrition information in a tabular or as provided below, linear (i.e., string) fashion rather than in vertical columns if the product has a total surface area available to bear labeling of less than 12 square inches, or if the product has a total surface area available to bear labeling of 40 or less square inches and the package shape or size cannot accommodate a standard vertical column or tabular display on any label panel. Nutrition information may be given in a linear fashion only if the label will not accommodate a tabular display.

(1) The following sample label illustrates the tabular display.

Nutrition Facts	Amount/serving	%DV*	Amount/serving	%DV*
	Total Fat 2g	**3%**	**Total Carb.** 0g	**0%**
Serving Size 1/3 cup (56g) Servings about 3	Sat. Fat 1g	**5%**	Fiber 0g	**0%**
Calories 90 Fat Cal. 20	*Trans* Fat 0.5g		Sugars 0g	
	Cholest. 10mg	**3%**	**Protein** 17g	
* Percent Daily Values (DV) are based on a 2,000 calorie diet.	**Sodium** 200mg	**8%**		
	Vitamin A 0% ● Vitamin C 0% ● Calcium 0% ● Iron 6%			

(2) The following sample label illustrates the linear format. When nutrition information is given in a linear fashion, bolding is required only on the title *"Nutrition Facts"* and is allowed voluntarily for the nutrient names for *"Calories," "Total fat," "Cholesterol," "Sodium," "Total carbohydrate,"* and *"Protein."*

Nutrition Facts Serv. Size: 1 package, Amount per Serving: **Calories** 45, Fat Cal. 10, **Total fat** 1g (2% DV), Sat. Fat 0.5g (3% DV), Trans Fat 0.5g, **Cholest.** 0mg (0% DV), **Sodium** 50mg (2% DV), **Total Carb.** 8g (3% DV), Fiber 1g (4% DV), Sugars 4g, **Protein** 1g, Vitamin A (8% DV), Vitamin C (8% DV), Calcium (0% DV), Iron (2% DV). Percent Daily Values (DV) are based on a 2,000 calorie diet.

(B) Using any of the following abbreviations:

Serving size - Serv size

Servings per container - Servings

Calories from fat - Fat cal

Calories from saturated fat - Sat fat cal

Saturated fat - Sat fat

Monounsaturated fat - Monounsat fat

Polyunsaturated fat - Polyunsat fat

Cholesterol - Cholest

Total carbohydrate - Total carb

Dietary fiber - Fiber

Soluble fiber - Sol fiber

Insoluble fiber - Insol fiber

Sugar alcohol - Sugar alc

Other carbohydrate - Other carb

(C) Omitting the footnote required in paragraph (d)(9) of this section and placing another asterisk at the bottom of the label followed by the statement *"Percent Daily Values are based on a 2,000 calorie diet"* and, if the term *"Daily Value"* is not spelled out in the heading, a statement that *"DV"* represents *"Daily Value."*

(D) Presenting the required nutrition information on any label panel.

(14) Shell eggs packaged in a carton that has a top lid designed to conform to the shape of the eggs are exempt from outer carton label requirements where the required nutrition information is clearly presented immediately beneath the carton lid or in an insert that can be clearly seen when the carton is opened.

(15) The unit containers in a multiunit retail food package where:

  (i)   The multiunit retail food package labeling contains all nutrition information in accordance with the requirements of this section;

  (ii) The unit containers are securely enclosed within and not intended to be separated from the retail package under conditions of retail sale; and

  (iii) Each unit container is labeled with the statement *"This Unit Not Labeled For Retail Sale"* in type size not less than 1/16 inch in height, except that this statement shall not be required when the inner unit containers bear no labeling at all. The word "*individual*" may be used in lieu of or immediately preceding the word "*Retail*" in the statement.

(16) Food products sold from bulk containers: *Provided,* That nutrition information required by this section be displayed to consumers either on the labeling of the bulk container plainly in view or in accordance with the provisions of paragraph (a)(2) of this section.

(17) Foods in packages that have a total surface area available to bear labeling greater than 40 square inches but whose principal display panel and information panel do not provide sufficient space to accommodate all required information may use any alternate panel that can be readily seen by consumers for the nutrition label. The space needed for vignettes, designs, and other nonmandatory label information on the principal display panel may be considered in determining the sufficiency of available space on the principal display panel for the placement of the nutrition label. Nonmandatory label information on the information panel shall not be considered in determining the sufficiency of available space for the placement of the nutrition label.

(18) Food products that are low-volume (that is, they meet the requirements for units sold in paragraphs (j)(18)(i) or (j)(18)(ii) of this section); that, except as provided in paragraph (j)(18)(iv) of this section, are the subject of a claim for an exemption that provides the information required under paragraph (j)(18)(iv) of this section, that is filed before the beginning of the time period for which the exemption is claimed, and that is filed by a person, whether it is the manufacturer, packer, or distributor, that qualifies to claim the exemption under the requirements for average full-time equivalent employees in paragraphs (j)(l8)(i) or (j)(18)(ii) of this section; and whose labels, labeling, and advertising do not provide nutrition information or make a nutrient content or health claim.

(i) For food products first introduced into interstate commerce before May 8, 1994, the product shall be exempt for the period:

(A) Between May 8, 1995, and May 7, 1996, if, for the period between May 8, 1994, and May 7, 1995, the person claiming the exemption employed fewer than an average of 300 full-time equivalent employees and fewer than 400,000 units of that product were sold in the United States; and

(B) Between May 8, 1996, and May 7, 1997, if for the period between May 8, 1995, and May 7, 1996, the person claiming the exemption employed fewer than an average of 200 full-time equivalent employees and fewer than 200,000 units of that product were sold in the United States.

(ii) For all other food products, the product shall be eligible for an exemption for any 12-month period if, for the preceding 12 months, the person claiming the exemption employed fewer than an average of 100 full-time equivalent employees and fewer than 100,000 units of that product were sold in the United States, or in the case of a food product that was not sold in the 12-month period preceding the period for which exemption is claimed, fewer than 100,000 units of such product are reasonably anticipated to be sold in the United States during the period for which exemption is claimed.

(iii) If a person claims an exemption under paragraphs (j)(18)(i) or (j)(18)(ii) of this section for a food product and then, during the period of such exemption, the number of full-time equivalent employees of such person exceeds the appropriate number, or the number of food products sold in the United States exceeds the appropriate number, or, if at the end of the period of such exemption, the food product no longer qualifies for an exemption under the provisions of paragraphs (j)(18)(i) or (j)(l8)(ii) of this section, such person shall have 18 months from the date that the product was no longer qualified as a low-volume product of a small business to comply with this section.

(iv) A notice shall be filed with the Office of Nutritional Products, Labeling, and Dietary Supplements (HFS-800), Center for Food Safety and Applied Nutrition, Food and Drug Administration, 5100 Paint Branch Pkwy., College Park, MD 20740 and contain the following information, except that if the person is not an importer and has fewer than 10 full-time equivalent employees, that person does not have to file a notice for any food product with annual sales of fewer than 10,000 total units:

(A) Name and address of person requesting exemption. This should include a telephone number or FAX number that can be used to contact the person along with the name of a specific contact;

(B) Names of the food products (including the various brand names) for which exemption is claimed;

(C) Name and address of the manufacturer, distributor, or importer of the food product for which an exemption is claimed, if different than the person that is claiming the exemption;

(D) The number of full-time equivalent employees. Provide the average number of full-time equivalent individuals employed by the person and its affiliates for the 12 months preceding the period for which a small business exemption is claimed for a product. The average number of full-time equivalent employees is to be determined by dividing the total number of hours of salary or wages paid to employees of the person and its affiliates by the number of hours of work in a year, 2,080 hours (i.e., 40 hours x 52 weeks);

(E) Approximate total number of units of the food product sold by the person in the United States in the 12-month period preceding that for which a small business exemption is claimed. Provide the approximate total number of units sold, or expected to be sold, in a 12-month period for each product for which an exemption is claimed. For products that have been in production for 1 year or more prior to the period for which exemption is claimed, the 12-month period is the period immediately preceding the period for which an exemption is claimed. For other products, the 12-month period is the period for which an exemption is claimed; and

(F) The notice shall be signed by a responsible individual for the person who can certify the accuracy of the information presented in the notice. The individual shall certify that the information contained in the notice is a complete and accurate statement of the average number of full-time equivalent employees of this person and its affiliates and of the number of units of the product for which an exemption is claimed sold by the person. The individual shall also state that should the average number of full-time equivalent employees or the number of units of food products sold in the United States by the person exceed the applicable numbers for the time period for which exemption is claimed, the person will notify FDA of that fact and the date on which the number of employees or the number of products sold exceeded the standard.

(v) FDA may by regulation lower the employee or units of food products requirements of paragraph (j)(18)(ii) of this section for any food product first introduced into interstate commerce after May 8, 2002, if the agency determines that the cost of compliance with such lower requirement will not place an undue burden on persons subject to it.

(vi) For the purpose of this paragraph the following definitions apply:

(A) *Unit* means the packaging or, if there is no packaging, the form in which a food product is offered for sale to consumers.

(B) *Food product* means food in any sized package which is manufactured by a single manufacturer or which bears the same brand name, which bears the same statement of identity, and which has similar preparation methods.

(C) *Person* means all domestic and foreign affiliates, as defined in *13 CFR 121.401*, of the corporation, in the case of a corporation, and all affiliates, as defined in *13 CFR 121.401*, of a firm or other entity, when referring to a firm or other entity that is not a corporation.

(D) *Full-time equivalent employee* means all individuals employed by the person claiming the exemption. This number shall be determined by dividing the total number of hours of salary or wages paid directly to employees of the person and of all of its affiliates by the number of hours of work in a year, 2,080 hours (i.e., 40 hours x 52 weeks).

Sec. 101.10 Nutrition labeling of restaurant foods

Nutrition labeling in accordance with Sec. 101.9 shall be provided upon request for any restaurant food or meal for which a nutrient content claim (as defined in Sec. 101.13 or in subpart D of this part) or a health claim (as defined in Sec. 101.14 and permitted by a regulation in subpart E of this part) is made, except that information on the nutrient amounts that are the basis for the claim (e.g., "low fat, this meal provides less than 10 grams of fat") may serve as the functional equivalent of complete nutrition information as described in Sec. 101.9. Nutrient levels may be determined by nutrient data bases, cookbooks, or analyses or by other reasonable bases that provide assurance that the food or meal meets the nutrient requirements for the claim. Presentation of nutrition labeling may be in various forms, including those provided in Sec. 101.45 and other reasonable means.

Adopted from 21 CFR Parts 101 through 169 (4/1/06 Edition)

17. Exemptions when it is Technologically Impracticable to Nutrition Label *(21 CFR 101.9(g)(9))*

(g)(9) When it is not technologically feasible, or some other circumstance makes it impracticable, for firms to comply with the requirements of this section (e.g., to develop adequate nutrient profiles to comply with the requirements of paragraph (c) of this section), FDA may permit alternative means of compliance or additional exemptions to deal with the situation. Firms in need of such special allowances shall make their request in writing to the Center for Food Safety and Applied Nutrition (HFS-800), Food and Drug Administration, 5100 Paint Branch Pkwy, College Park, MD 20740.

Food Labeling; Guidelines for Voluntary Nutrition Labeling of Raw Fruits, Vegetables, and Fish

The Food and Drug Administration (FDA) has amended the voluntary nutrition labeling regulations by updating the names and the nutrition labeling values for the 20 most frequently consumed raw fruits, vegetables, and fish in the United States and clarified guidelines for the voluntary nutrition labeling of these foods.

The updated nutrition labeling values in retail stores and on individually packaged raw fruits, vegetables, and fish will enable consumers to make better purchasing decisions to reflect their dietary needs.

The effective date for these guidelines is January 1, 2008.

Section 101.44 of these regulations is revised to read as follows:

What are the 20 most frequently consumed raw fruits, vegetables, and fish in the United States?

(a) The 20 most frequently consumed raw fruits are: Apple, avocado (California), banana, cantaloupe, grapefruit, grapes, honeydew melon, kiwifruit, lemon, lime, nectarine, orange, peach, pear, pineapple, plums, strawberries, sweet cherries, tangerine, and watermelon.

(b) The 20 most frequently consumed raw vegetables are: Asparagus, bell pepper, broccoli, carrot, cauliflower, celery, cucumber, green (snap) beans, green cabbage, green onion, iceberg lettuce, leaf lettuce, mushrooms, onion, potato, radishes, summer squash, sweet corn, sweet potato, and tomato.

(c) The 20 most frequently consumed raw fish are: Blue crab, catfish, clams, cod, flounder/sole, haddock, halibut, lobster, ocean perch, orange roughy, oysters, pollock, rainbow trout, rockfish, salmon (Atlantic/coho/Chinook/sockeye, chum/pink), scallops, shrimp, swordfish, tilapia, and tuna.

Section 101.45 is amended by revising paragraph (a)(3)(iii) and adding paragraph (a)(3)(iv) to read as follows:

What are the guidelines for the voluntary nutrition labeling of raw fruits, vegetables, and fish?

(a) ★ ★ ★

(3) ★ ★ ★

(iii) When retailers provide nutrition labeling information for more than one raw fruit or vegetable on signs or posters or in brochures, notebooks, or leaflets, the listings for saturated fat, *trans* fat, and cholesterol may be omitted from the charts or individual nutrition labels if a footnote states that most fruits and vegetables provide negligible amounts of these nutrients, but that avocados contain 0.5 gram (g) of saturated fat per ounce (e.g., "Most fruits and vegetables provide negligible amounts of saturated fat, *trans* fat, and cholesterol; avocados provide 0.5 g of saturated fat per ounce"). The footnote also may contain information about the polyunsaturated and monounsaturated fat content of avocados.

(iv) When retailers provide nutrition labeling information for more than one raw fish on signs or posters or in brochures, notebooks, or leaflets, the listings for *trans* fat, dietary fiber, and sugars may be omitted from the charts or individual nutrition labels if the following footnote is used, "Fish provide negligible amounts of *trans* fat, dietary fiber, and sugars."

Chapter VI

* * * * *

The appendices C and D to part 101 are also revised. Copies of the entire document may be obtained from the *Federal Register*: July 25, 2006 (Volume 71, Number 142), Page 42031- 42047, and from the Federal Register Online via GPO Access [wais.access.gpo.gov], [DOCID:fr25jy06-8.

Section F: Claims

1. **Nutrient Content Claims Authorized by Regulation (21 CFR 101.13)**

    (a) **Summary Comments**

    Food labels and labeling may also bear authorized nutrient content claims *(Sec. 403(r)(1)(A) of the FD&CA)*. A nutrient content claim is a claim characterizing the level of a nutrient in a food. FDA authorizes nutrient content claims following the submission of a petition and promulgation of a regulation or through a notification procedure for a claim based on an authoritative statement. A food bearing a nutrient content claim that has not been authorized by regulation or by the Act misbrands the product *(Sec. 403(r)(1)(A) of the FD&CA)*.

    A manufacturer may make a statement about a substance for which there is no established RDI or DRV so long as the claim specifies only the amount of the substance per serving and does not imply that there is a lot or a little of the substance in the product, e.g., "0.3 g omega 3 fatty acids."

    The nutrient content claims that FDA has authorized by regulation are included in this manual in Chart VI.F.1(c). Most nutrient content claims, including "*high*" and "*more,*" are defined only for substances with an established Reference Daily Intake (RDI) or Daily Reference Value (DRV). A list of nutrients with RDIs and DRVs can be found in Chart VI.E.12.

    (b) **Questions and Responses**

    The questions and responses sections included under Chapter VI are a series of established requirements designed to aid the reviewers in establishing the degree of compliance that a specific food label and its labeling comply with applicable laws and regulations. It is not intended to represent all questions that may arise during the review process, but as examples of typical questions during a label review. This section can also serve as a teaching aid for less experienced reviewers and as a ready reference for experienced reviewers. It can also serve as an example for developing a response to advise those responsible for food labels and labeling of needed corrections.

Is the product intended for human food and offered for sale?

   A. If "*YES,*" continue.
   B. If "*NO,*" STOP HERE. Nutrient content claims regulations are not applicable.

Is the product a restaurant food or meal that bears a nutrient content claim in accordance with 21 CFR 101.13 (Nutrient content claims–general principles) and 21 CFR Part 101, Subpart D (Specific requirements for nutrient content claims), as required by 21 CFR 101.10)?

   A. If "*YES,*" continue.
   B. If "*NO,*" because the product is not a restaurant food or a meal, or if it is a restaurant food or a meal and it bears a nutrient content claim, continue.

      If "*NO,*" because the product is a restaurant food or meal and bears a nutrient content claim, but fails to comply with the requirement(s) of *21 CFR 101.10*, state that the food is a restaurant food or meal (whichever is applicable) that bears the nutrient content claim "_____" (insert specific nutrient content claim declared on the label, e.g., "low fat"), but _____ (insert the deviation from the specific requirement, e.g., contains more than the 3g of fat per reference amount customarily consumed in a food bearing this claim), as provided for in _____ (Insert CFR reference, e.g., *21 CFR 101.62(b)(2)(i)(A))*.

Chapter VI

Q *Does the label or labeling of a food that is not a restaurant food, or meal that bears a claim that expressly or implicitly characterizes the level of a nutrient (nutrient content claim) of the type required in nutrition labeling under 21 CFR 101.9, conform to the applicable requirements in 21 CFR 101.13 and in 21 CFR Part 101, Subpart D?*

A. If "*YES*," continue.

B. If "*NO*," because the label fails to comply with the requirements for a specific nutrient claim, state that the label or labeling bears the nutrient content claim "_____" (insert the specific nutrient content claim declared on the label), which is of the type provided for under *21 CFR 101*.9, but the claim fails to conform to requirement(s) of _____ (insert the section of the regulation from which the claim deviates) in that it _____ (insert the specific deviation.)

[NOTE: Information that is required or permitted by 21 CFR 101.9 to be declared in nutrition labeling, but is not a nutrient content claim, is not subject to the nutrient content claim requirements and general principles. However, when such information is declared elsewhere on the label, it is a nutrient content claim and is subject to the nutrient content claim requirements in 21 CFR 101.13(c).]

Q *Is the product a substitute food?*

[NOTE: A "substitute" food, as used here, is one that may be used interchangeably with another food that it resembles (i.e., it is organoleptically, physically, and functionally similar, including shelf life), but is not nutritionally inferior to the food it resembles, unless the food is labeled as an "imitation".]

A. If "*YES*," the difference in performance characteristics that materially limits the use of the food must be included in a disclaimer adjacent to the most prominent claim, as defined in *21 CFR 101.13(j)(2)(iii)*, e.g., "not recommended for frying", as required by *21 CFR 101.13(d)(1)*.

[NOTE: The disclaimer must be in easily legible print or type and in a size not less than that required by 21 CFR 101.105(i), except where the size of the claim is less than two times the required size of the net quantity of contents statement as further provided under 21 CFR 101.13(d)(2).]

B. If "*NO*," continue.

Q *Does the label of a product bearing a "free" or "low" claim before the food name imply that the food differs from other foods of the same type by having a lower amount of a nutrient as a result of having been specially processed, altered, or reformulated to lower the amount of the nutrient in the food, or to remove the nutrient from the food, or by not including the nutrient in the food as provided for by 21 CFR 101.13(e)(1)?*

A. If "*YES*," it may bear a claim, e.g., low sodium potato chips, continue.

B. If "*NO*," because the product bears a claim for which it qualifies, but has not been specially processed, altered, formulated, or reformulated to qualify for that claim, state that the label deviates from the requirements of *21 CFR 101.13(e)(2)*, in that it fails to indicate that the food inherently meets the claim and that the claim clearly refers to all foods of that type and not merely to the particular brand to which the labeling is attached.

Is the nutrient content claim declared in a type size no larger than two times that of the statement of identity and that is not in an unduly prominent type style compared to the statement of identity? *(21 CFR 101.13(f))*

A. If "*YES*," continue.
B. If "*NO*," state that the nutrient content claim "_____" (insert the specific claim made on the label) is not declared in _____ (insert the specific deviation(s) from the requirements), as required by *21 CFR 101.13(f).*

 Does the label of the food bear a disclosure statement in accordance with the following requirements? *(See Regulations VI.F.1(d))*

[**NOTE:** *Food products that bear nutrient content claims, but contain more than 13.0g of fat, 4.0g of saturated fat, 60mg of cholesterol, or 480mg of sodium per reference amount customarily consumed, must bear a disclosure statement.]*

A. If "*YES*," continue.
B. If "*NO*," because the product does not exceed these levels, continue.

If "*NO*," because the label fails to bear the required disclosure statement, state that the label bears the nutrient content claim(s) "_____" (specify the nutrient content claim on the label) and the nutrient level for "_____" (specify nutrient(s)) exceeds "_____" (insert the designated level for the particular nutrient) per labeled serving, but the label fails to disclose that the nutrient exceeding the specified level is present in the food, in accordance with *21 CFR 101.9(h)(1), (2),* or *(3),* as appropriate. (Select applicable reference.)

 Does the label bear a statement about the amount or percentage of a nutrient in the product, in accordance with the following requirements, as provided in 21 CFR 101.13(i)?

The label may bear a statement of the amount or percentage of a nutrient if:
1. *The use of an amount or percentage statement on the label of a food implicitly characterizes the level of the nutrient in the food, and is consistent with a definition for a claim, as provided in "Subpart D – Specific Requirements for Nutrient Content Claims" (21 CFR Part 101, Subpart D). Such a claim might be, "less than 3g of fat per serving;"*
2. *The use of statement on the food implicitly characterizes the level of the nutrient in the food and is not consistent with such a definition, but the label carries a disclaimer adjacent to the statement that the food is not "low" in or a "good source" of the nutrient, such as "only 200mg sodium per serving, not a low sodium food," in accordance with the type size specified in 21 CFR 101.13(i)(2); or*
3. *The statement does not in any way implicitly characterize the level of the nutrient in the food, and it is not false or misleading in any respect (e.g., "5 grams of fat"), in which case no disclaimer is required.*
4. *"Percent fat free" claims are not authorized by 21 CFR 101.13)(i). Such claims shall comply with 21 CFR 101.62(b)(6).*

A. If "*YES*," continue.
B. If "*NO*," because such claims are made, continue.

If "*NO*," because the label fails to comply with a specific requirement under this section, state that the label bears the implicitly characterizing statement "_____" (insert the specific characterizing statement appearing on the label), but fails to comply with the requirement that _____ (insert deviation from regulation, including CFR reference).

Chapter VI

Q *Does the food bear a statement that compares the level of a nutrient in the food with the level of a nutrient in a reference food (e.g., "light", "reduced", "less", "fewer", or "more" claims) in accordance with 21 CFR 101.13(j)(1) and (j)(2)?*

1. *The amount of that nutrient in the food must be compared to an amount of the nutrient in an appropriate reference food as specified below:*

   a)(i) *For "less" (or "fewer") and "more" claims, the reference food may be a dissimilar food within a product category that can generally be substituted for one another in the diet (e.g., potato chips as a reference for pretzels, orange juice as reference for vitamin C tablets) or a similar food (e.g., potato chips as reference for potato chips, one brand of multivitamin for another brand of multivitamin).*

   b) *For "light," "reduced," "added," "extra," "fortified," and "enriched" claims, the reference food must be a similar food (e.g., potato chips as a reference for potato chips, one brand of multivitamin for another brand of multivitamin).*

   a)(i) *For "light" claims, the reference food must be representative of the type of food that includes the product that bears the claim. The nutrient value for the reference food must be representative of a broad base of foods of that type, e.g., a value in a representative, valid data base; an average value determined from the top three national (or regional) brands, a market basket norm, or, where its nutrient value is representative of the food type, a market leader. Firms using such a reference nutrient value as a basis for a claim are required to provide specific information upon which the nutrient value was derived, on request, to consumers and appropriate regulatory officials.*

   b) *For relative claims other than "light," including "less" and "more" claims, the reference food may be the same as that provided for "light" in paragraph (b)(i) above, or it may be the manufacturer's regular product, or that of another manufacturer, that has been offered for sale to the public on a regular basis for a substantial period of time in the same geographic area by the same business entity or by one entitled to use its trade name.*

2. *Foods bearing relative claims:*

   a) *The label or labeling must state the identity of the reference food and the percentage (or fraction) of the amount of the nutrient in the reference food by which the nutrient in the labeled food differs (e.g., "50 percent less fat than (reference food)" or "1/3 fewer calories than (reference food)").*

   b) *The information on the reference food must be immediately adjacent to the most prominent claim. The type size shall be in accordance with 21 CFR 101.13(g)(1).*

   c) *The determination of which use of the claim is in the most prominent location on the label or labeling will be made based on the following factors, considered in order.*
      (i) A claim on the principal display panel adjacent to the statement of identity;
      (ii) A claim elsewhere on the principal display panel;
      (iii) A claim on the information panel; or
      (iv) A claim elsewhere on the label or labeling.

   d) *The label or labeling must also bear:*
      (i) Clear and concise quantitative information comparing the amount of the subject nutrient in the product per labeled serving with that in the reference food; and
      (ii) This statement that must appear adjacent to the most prominent claim or to the nutrition label, except that if the nutrition label is on the information panel, the quantitative information may be located elsewhere on the information panel in accordance with 21 CFR 101.2.

3. *A relative claim for decreased levels of a nutrient may not be made on the label or in labeling of a food if the nutrient content of the reference food meets the requirement for a "low" claim for that nutrient (e.g., 3g fat or less).*

A. If "*YES*," continue.

B. If "*NO*," because no claims are made, continue.

If "*NO*," that the product makes a relative claim but fails to comply with the above requirements, state that the _____ (insert the location of the claim, e.g., the principal display panel of the label, or the labeling such as brochure) makes the relative claim "_____" (insert the specific claim made), but fails to bear the "_____" (insert deviations from specific requirements), in accordance with _____ (insert CFR reference).

Does the identity statement bear the term "modified" for a food that bears a relative claim and complies with the requirements listed below? (21 CFR 101.13(k))

1. *The term "modified" is followed immediately by the name of the nutrient whose content has been altered (e.g., "modified fat cheesecake").*
2. *The statement of identity must be immediately followed by a comparative statement, such as "contains 35 percent less fat than _____."*
3. *The label or labeling must also bear the information required by 21 CFR 101.13(j)(2), regarding relative claims.*

A. If "*YES*," continue.

B. If "*NO*," because the label fails to bear such term, continue.

If "*NO*," because the identity statement bears the term "*modified*" and a relative claim provided for by *21 CFR 101.13(k)*, that is not in accordance with the requirement(s) in *21 CFR 101.13(k)*, state that the label bears the term "*modified*," and the relative claim "_____" (insert the specific relative claim as declared on the label), but fails to bear _____ (insert the applicable required statement and where it should be declared, e.g., the name of the food immediately followed by the name of the nutrient whose content has been altered e.g., "Modified fat cheesecake"), as required by *21 CFR 101.13(k)*.

Does a "meal product" bearing claims comply with the requirements of 21 CFR 101.13(l)?

1. *The meal product makes a major contribution to the diet by:*
   a) *Weighing at least 10 ounces per labeled serving; and*
   b) *Containing not less than three 40-gram portions of food, or combinations of foods, from two or more of the following four food groups, except as noted in item (v) below:*
      (i) *Bread, cereal, rice, and pasta group;*
      (ii) *Fruits and vegetables group;*
      (iii) *Milk, yogurt, and cheese group;*
      (iv) *Meat, poultry, fish, dry beans, eggs, and nuts group; except that;*
      (v) *These foods must not be sauces (except for foods in the above four food groups that are in the sauces), gravies, condiments, relishes, pickles, olives, jams, jellies, syrups, breading, or garnishes; and*
2. *The meal product is represented as, or is in a form commonly understood to be, a breakfast, lunch, dinner, or meal. Such representations may be made by statements, photographs, or vignettes. (21 CFR 101.13(l))*

A. If "*YES*," continue.

B. If "*NO*," because it fails to comply with the requirements under *21 CFR 101.13(l)*, state that the product is represented to be a meal and bears the claim "_____" (insert the claim made on the label or labeling), but it fails to comply with the requirement

_____ (specify deviation from the specific regulations and include CFR reference).

Q *Does the food of a "main dish product" bearing claims comply with the requirements of 21 CFR 101.13(m) as follows?*

1. *The main dish product makes a major contribution to a meal by:*
   a) *Weighing at least 6 ounces per labeled serving; and*
   b) *Containing not less than 40 grams of food, or combinations of foods, from each of at least two of the following four food groups, except as noted in item (v) below:*
      (i) *Bread, cereal, rice, and pasta group;*
      (ii) *Fruits and vegetables group;*
      (iii) *Milk, yogurt, and cheese group;*
      (iv) *Meat, poultry, fish, dry beans, eggs, and nuts group; except that;*
      (v) *These foods must not be sauces (except for foods in the above four food groups that are in the sauces), gravies, condiments, relishes, pickles, olives, jams, jellies, syrups, breading, or garnishes; and*
2. *The main dish product is represented as, or is in a form commonly understood to be, a main dish (e.g., not a beverage or dessert). Such representations may be made by statements, photographs, or vignettes. (21 CFR 101.13(m))*

A. If "*YES,*" continue.
B. If "*NO,*" because the label bears a claim, but the product fails to comply with the requirements of *21 CFR 101.13(m)*, state that the label bears the claim "_____" (insert the claim made on the label or labeling), but it fails to comply with _____ (specify deviation from the specific regulations and include CFR reference).

Q *Does the label that bears a term such as "healthy," "health," "healthful," "healthfully," "healthfulness," "healthily," "healthier," "healthiest," "or "healthiness" (as an implied nutrient content claim) meet the requirements of 21 CFR 101.65(d)?* *(See Chart VI.F.1(e))*

A. If "*YES,*" continue.
B. If "*NO,*" state that the label bears the implied claim "_____" (insert claim declared on the label, e.g., "healthy"), but that the food fails to _____ (insert specific requirements that the food fails to meet, e.g., meet the definition for low fat), as required by _____ (insert CFR reference, e.g., *21 CFR 101.65(d)(2)(i)*).

Chart VI.F.1(c)
Definitions of Nutrient Content Claims
(21 CFR 101)

Nutrients	Free	Low	Reduced/Less	Comments
	• Synonyms for *"Free"*: *"Zero," "No," "Without," "Trivial Source of," "Negligible Source of," "Dietarily Insignificant Source of."* • Definitions for *"Free"* for meals and main dishes are the stated values per labeled serving.	• Synonyms for *"Low,"*: *"Little,"* (*"Few"* for Calories), *"Contains a Small Amount of," "Low Source of."*	• Synonyms for *"Reduced"/"Less"*: *"Lower"* (*"Fewer"* for calories). *"Modified"* may be used in statement of identity. • Definitions for meals and main dishes are same as for individual foods on a per 100g basis.	• For *"Free," "Very Low,"* or *"Low"* must indicate if food meets a definition, without benefit of special processing, alteration, formulation or reformulation; e.g., *"broccoli, a fat free food,"* or *"celery, a low calorie food."* Nonfat is an approved synonym for *"Fat-free."*
Calories § *101.60(b)*	• Less than 5 cal per reference amount and per labeled serving. • Not defined for meals or main dishes.	• 40 cal or less per reference amount (and per 50g if reference amount is small). • Meals and main dishes: 120 cal or less per 100g.	• At least 25% fewer calories per reference amount than an appropriate reference food. • Reference food may not be *"Low Calorie."* • Use term *"Fewer"* rather than *"Less."*	• *"Light"* or *"Lite"*: If 50% or more of the calories are from fat, fat must be reduced by at least 50% per reference amount. If less than 50% of calories are from fat, fat must be reduced at least 50% or calories reduced at least $\frac{1}{3}$ per reference amount. • *"Light"* or *"Lite"* meal or main dish product meets definition for *"Low Calorie"* or *"Low Fat"* meal and is labeled to indicate which definition is met.

Adopted from FDA Guide to Food Labeling Act (NLEA) Requirements,
August 1999)

Chart VI.F.1(c)
(Continued)

Nutrients	Free	Low	Reduced/Less	Comments
				▪ For dietary supplements: calorie claims can only be made when the reference product contains more than 40 calories per serving.
Total Fat § 101.62(b)	▪ Less than 0.5g per reference amount and per labeled serving (or for meals and main dishes, less than 0.5g per labeled serving). ▪ No ingredient that is fat or understood to contain fat except as noted below.[1]	▪ 3g or less per reference amount (and per 50g if reference amount is small). ▪ Meals and main dishes: 3g or less per 100g and not more than 30% of calories from fat.	▪ At least 25% less fat per reference amount than an appropriate reference food. ▪ Reference food may not be *"Low Fat."*	▪ *"__ % Fat Free"*: OK if food meets the requirements for *"Low Fat."* ▪ 100% Fat Free: food must be *"Fat Free."* ▪ *"Light"* see above. ▪ For dietary supplements: fat claims cannot be made for products that contain 40 calories or less per serving.

Chart VI.F.1(c)
(Continued)

Chapter VI

Nutrients	Free	Low	Reduced/Less	Comments
Saturated Fat § *101.62(c)*	▪ Less than 0.5g saturated fat and less than 0.5g *trans* fatty acids per reference amount and per labeled serving (or for meals and main dishes, less than 0.5g saturated fat and less than 0.5g *trans* fatty acids per labeled serving). ▪ No ingredient that is understood to contain saturated fat except as noted below.[1]	▪ 1g or less per reference amount and 15% or less calories from saturated fat. ▪ Meals and main dishes: 1g or less per 100g and less than 10% of calories from saturated fat.	▪ At least 25% less saturated fat per reference amount than an appropriate reference food. ▪ Reference food may not be *"Low Saturated Fat"*	▪ Next to all saturated fat claims, must declare the amount of cholesterol if 2mg or more per reference amount; and the amount of total fat if more than 3g per reference amount (0.5g or more of total fat for *"Saturated Fat Free"*). ▪ For dietary supplements: saturated fat claims cannot be made for products that contain 40 calories or less per serving.

Adopted from FDA Guide to Food Labeling Act (NLEA) Requirements, August 1999

Nutrients	Free	Low	Reduced/Less	Comments
Cholesterol § *101.62(d)*	▪ Less than 2mg per reference amount and per labeled serving (or for meals and main dishes, less than 2mg per labeled serving). No ingredients that contain cholesterol except as noted below.[1] ▪ If less than 2mg per reference amount by special processing and total fat exceeds 13g per reference amount and labeled serving, the amount of cholesterol must be *"Substantially Less"* (25%) than in a reference food with significant market share (5% of market).	▪ 20mg or less per reference amount (and per 50g of food if reference amount is small). ▪ If qualifies by special processing and total fat exceeds 13g per reference amount and labeled serving, the amount of cholesterol must be *"Substantially Less"* (25%) than in a reference food with significant market share (5% of market). ▪ Meals and main dishes: 20mg or less per 100g.	▪ At least 25% less cholesterol per reference amount than an appropriate reference food. ▪ Reference food may not be *"Low Cholesterol."*	▪ Cholesterol claims only allowed when food contains 2g or less saturated fat per reference amount, or for meals and main dish products per labeled serving size for *"free"* claims or per 100g for *"low"* and *"reduced/less"* claims. ▪ Must declare the amount of total fat next to cholesterol claim when fat exceeds 13g per reference amount and labeled serving (or per 50g of food if reference amount is small), or when the fat exceeds 19.5g per labeled serving for main dishes or 26g for meal products. ▪ For dietary supplements: cholesterol claims cannot be made for products that contain 40 calories or less per serving.

Chart VI.F.1(c)
(Continued)

Chapter VI

Nutrients	Free	Low	Reduced/Less	Comments
Sodium § 101.61	▪ Less than 5mg per reference amount and per labeled serving (or for meals and main dishes, less than 5mg per labeled serving). ▪ No ingredient that is sodium chloride or generally understood to contain sodium except as noted below.[1]	▪ 140mg or less per reference amount (and per 50g if reference amount is small). ▪ Meals and main dishes: 140mg or less per 100g.	▪ At least 25% less sodium per reference amount than an appropriate reference food. ▪ Reference food may not be *"Low Sodium."*	▪ *"Light"* (for sodium reduced products): If food is *"Low Calorie"* and *"Low Fat"* and sodium is reduced by at least 50%. ▪ *"Light in Sodium"*: If sodium is reduced by at least 50% per reference amount. Entire term *"Light in Sodium"* must be used in the same type size, color & prominence. Light in Sodium for meals = *"Low in Sodium."* ▪ *"Very Low Sodium"*: 35mg or less per reference amount (and per 50g if reference amount is small). For meals and main dishes: 35mg or less per 100g. ▪ *"Salt Free"* must meet criterion for *"Sodium Free"*. ▪ *"No Salt Added"* and *"Unsalted"* must meet conditions of use and must declare *"This is Not A Sodium Free Food"* on information

Adopted from FDA Guide to Food Labeling Act (NLEA) Requirements,
August 1999)

Chart VI.F.1(c)
(Continued)

Nutrients	Free	Low	Reduced/Less	Comments
				▪ panel if food is not *"Sodium Free."* ▪ *"Lightly Salted"*: 50% less sodium than normally added to reference food and if not *"Low Sodium"* so labeled on information panel.
Sugars § *101.60(c)*	▪ *"Sugar Free"*: Less than 0.5g sugars per reference amount and per labeled serving (or for meals and main dishes, less than 0.5g per labeled serving). ▪ No ingredient that is a sugar or generally understood to contain sugars except as noted below.* ▪ Disclose calorie profile (e.g., *"Low Calorie"*).	▪ Not defined. No basis for a recommended intake.	▪ At least 25% less sugars per reference amount than an appropriate reference food. ▪ May not use this claim on dietary supplements of vitamins and minerals.	▪ *"No Added Sugars"* and *"Without Added Sugars"* are allowed if no sugar or sugar-containing ingredient is added during processing. State if food is not *"Low"* or *"Reduced Calorie."* ▪ The terms *"Unsweetened"* and *"No Added Sweeteners"* remain as factual statements. ▪ Claims about reducing dental cavities are implied health claims. ▪ The term "Sugars" does not include sugar alcohols.

Adopted from FDA Guide to Food Labeling Act (NLEA) Requirements, August 1999)

Chart VI.F.1(c)
(Continued)

Notes

[1]   Except if the ingredient listed in the ingredient statement has an asterisk that refers to a footnote (e.g., *"adds a trivial amount of fat"*).

"Reference Amount" = reference amount customarily consumed.

"Small Reference Amount" = reference amount of 30g or less or 2 tablespoons or less (for dehydrated foods that are typically consumed when reconstituted with water or a diluent containing an insignificant amount, as defined in §101.9(f)(1), of all nutrients per reference amount, the per 50g criterion refers to the prepared form of the food).

- *When levels exceed: 13 g Fat, 4 g Saturated Fat, 60 mg Cholesterol, and 480 mg Sodium per reference amount, per labeled serving or, for foods with small reference amounts, per 50 g, a disclosure statement is required as part of claim (e.g., "See nutrition information for___content" with the blank filled in with nutrient(s) that exceed the prescribed levels).*

Relative (or Comparative) Claims Accompanying Information

For all relative claims, percent (or fraction) of change and identity of reference food must be declared in immediate proximity to the most prominent claim. Quantitative comparison of the amount of the nutrient in the product per labeled serving with that in reference food must be declared on information panel.

For "Light" claims: Generally, percentage reduction for both fat and calories must be stated. An exception is that percentage reduction need not be specified for "low-fat" products. Quantitative comparisons must be stated for both fat and calories.

For claims characterizing the level of antioxidant nutrients in a food:

- an RDI must be established for each of the nutrients that are the subject of the claim;

- each nutrient must have existing scientific evidence of antioxidant activity and

- the level of each nutrient must be sufficient to meet the definition for "high," "good source," or "high potency" in *21 CFR 101.54(b),(c), or (e).*

Beta-carotene may be the subject of an antioxidant claim when the level of vitamin A present as beta-carotene in the food is sufficient to qualify for the claim.

Adopted from FDA Guide to Food Labeling Act (NLEA) Requirements, August 1999)

Chart VI.F.1(c)
(Continued)

Reference Food	
"Light" or "Lite"	(1) A food representative of the type of food bearing the claim (e.g., average value of top three brands or representative value from valid data base), (2) Similar food (e.g., potato chips for potato chips), and (3) Not low-calorie <u>and</u> low-fat (except light-sodium foods which <u>must</u> be low-calorie & low-fat).
"Reduced" and "Added" (or "Fortified" and "Enriched")	(1) An established regular product or average representative product, and (2) Similar food.
"More" and "Less" (or "Fewer")	(1) An established regular product or average representative product, and (2) A dissimilar food in the same product category which may be generally substituted for the labeled food (e.g., potato chips for pretzels) or a similar food.

Other Nutrient Content Claims	
"Lean"	On seafood or game meat that contains less than 10g total fat, 4.5g or less saturated fat, and less than 95mg cholesterol per reference amount and per 100g (for meals & main dishes, meets criteria per 100g and per labeled serving).
"Extra Lean"	On seafood or game meat that contains less than 5g total fat, less than 2g saturated fat and less than 95mg cholesterol per reference amount and per 100g (for meals and main dishes, meets criteria per 100g and per labeled serving).
"High Potency"	May be used on foods to describe individual vitamins or minerals that are present at 100% or more of the RDI per reference amount or on a multi-ingredient food product that contains 100% or more of the RDI for at least 2/3 of the vitamins and minerals with DV's and that are present in the product at 2% or more of the RDI (e.g., "High potency multivitamin, multimineral dietary supplement tablets").
"High", "Rich In", or "Excellent Source Of"	Contains 20% or more of the Daily Value (DV) to describe protein, vitamins, minerals, dietary fiber, or potassium per reference amount. May be used on meals or main dishes to indicate that product contains a food that meets definition. May not be used for total carbohydrate.
"Good Source of", "Contains" or "Provides"	10%-19% of the DV per reference amount. These terms may be used on meals or main dishes to indicate that product contains a food that meets definition. May not be used for total carbohydrate.
"High Potency"	May be used on foods to describe individual vitamins or minerals that are present at 100% or more of the RDI per reference amount or on a multi-ingredient food product that contains 100% or more of the RDI for at least 2/3 of the vitamins and minerals with DV's and that are present in the product at 2% or more of the RDI (e.g., "High potency multivitamin, multimineral dietary supplement tablets").
"High", "Rich In", or "Excellent Source Of"	Contains 20% or more of the Daily Value (DV) to describe protein, vitamins, minerals, dietary fiber, or potassium per reference amount. May be used on meals or main dishes to indicate that product contains a food that meets definition. May not be used for total carbohydrate.
"Good Source of", "Contains" or "Provides"	10%-19% of the DV per reference amount. These terms may be used on meals or main dishes to indicate that product contains a food that meets definition. May not be used for total carbohydrate.
"More", "Added", "Extra", or "Plus"	10% or more of the DV per reference amount. May only be used for vitamins, minerals, protein, dietary fiber, and potassium.
"Modified"	May be used in statement of identity that bears a relative claim (e.g., "Modified Fat Cheese Cake, contains 35% Less Fat than our Regular Cheese Cake.").
Any Fiber Claim	If food is not low in total fat, must state total fat in conjunction with claim such as "More Fiber".

Adopted from FDA Guide to Food Labeling Act (NLEA) Requirements, August 1999)

Chart VI.F.1(c)	Chapter VI
(Continued)	

Implied Claims

- Claims about a food or ingredient that suggests that the nutrient or ingredient are absent or present in a certain amount or claims about a food that suggests a food may be useful in maintaining healthy dietary practices and which are made with an explicit claim (e.g. "healthy, contains 3 grams of fat") are implied claims and are prohibited unless provided for in a regulation by FDA. In addition, the Agency has devised a petition system whereby specific additional claims may be considered.

- Claims that a food contains or is made with an ingredient that is known to contain a particular nutrient may be made if product is "Low" in or a "Good Source" of the nutrient associated with the claim (e.g. "good source of oat bran").

- Equivalence claims: "contains as much [nutrient] as a [food]" may be made if both reference food and labeled food are a "Good Source" of a nutrient on a per serving basis. (e.g. "Contains as much vitamin C as an 8 ounce glass of orange juice").

- The following label statements are generally not considered implied claims unless they are made in a nutrition context: 1) avoidance claims for religious, food intolerance, or other non-nutrition related reasons (e.g. "100% milk free"); 2) statements about non-nutritive substances (e.g. "no artificial colors"); 3) added value statements (e.g. "made with real butter"); 4) statements of identity (e.g. "corn oil" or "corn oil margarine"); and 5) special dietary statements made in compliance with a specific Part 105 provision.

Claims on Foods for Infants and Children Less than 2 Years of Age

Nutrient content claims are not permitted on foods intended specifically for infants and children less than 2 years of age except:

1. Claims describing the percentage of vitamins and minerals in a food in relation to a daily value.

2. Claims on infant formulas provided for in Part 107.

3. The terms "Unsweetened" and "Unsalted" as taste claims.

4. "Sugar Free" and "No Added Sugar" claims on dietary supplements only.

Terms Covered That Are Not Nutrient Content Claims *(See 21 CFR 101.95)*	
"Fresh"	A raw food that has not been frozen, heat processed, or otherwise preserved.
"Fresh Frozen"	Food was quickly frozen while still fresh.

Chapter VI

Chart VI.F.1(d)
Disclosure Statement and Criteria
(21 CFR 101.13(h))

Disclosure Statement

Some products that are permitted to bear nutrient content claims contain certain other nutrients that increase the risk of disease or health-related conditions, which are diet related. The labels of these products must bear a disclosure statement, such as "See nutrition information for fat content," to highlight the nutrient that exceeds the disclosure level specified in the chart below. Such disclosure statement must be immediately adjacent to the nutrient content claim in easily legible boldface print or typ*e (21 CFR 101.13(h)(4))*.   Alternatively, the label may disclose the amount of the specified nutrient adjacent to the claim, e.g., "High in fiber, Contains 5g of fat. See side panel for nutrition information."

Several circumstances in which a disclosure statement is required include the following:
1) If a claim is made about fiber on a product that is not low fat;
2) If a claim is made about cholesterol on a product that exceeds the disclosure level of fat;
3) If a "free" claim is made about saturated fat on a product that is not cholesterol and fat free (i.e., less than 0.2mg cholesterol and less than 0.5g fat); and
4) If a low or reduced saturated fat claim is made on a product that is not cholesterol free and low fat (i.e., 3g fat or less per reference amount).

Disclosure levels for sodium, fat, saturated fat, and cholesterol vary with the types of food, i.e., individual foods, meals, and main dishes.

DISCLOSURE STATEMENTS CHART			
Nutrient	**Individual Food** *	**Meals****	**Main Dishes** **
Fat	13.0 g	26 g	19.5 g
Saturated fat	4.0 g	8.0 g	6.0 g
Cholesterol	60 mg	120 mg	90 mg
Sodium	480 mg	960 mg	720 mg

*   Per reference amount, per labeled serving and, if the reference amount is small (i.e., 30 g or less or 2 tbsp or less), per 50 g.
**   Per labeled serving *(§101.13(h))*.

Chart VI.F.1(e)
Conditions for
Use of "Healthy"

Conditions for Use of "Healthy":

The FDA has revised the regulatory text of the "healthy" regulation to clarify the scope and meaning of the regulation and to reformat the nutrient content requirements for "healthy" into a more readable set of tables, consistent with the Presidential Memorandum instructing that regulations be written in plain English.

Section 101.65 is amended by revising paragraph (d) to read as follows:

Sec. 101.65 Implied nutrient content claims and related label statements.
*****

(d) General nutritional claims. (1) This paragraph covers labeling claims that are implied nutrient content claims because they:
  (i) Suggest that a food because of its nutrient content may help consumers maintain healthy dietary practices; and
  (ii) Are made in connection with an explicit or implicit claim or statement about a nutrient (e.g., "healthy, contains 3 grams fat").
(2) You may use the term ``healthy'' or related terms (e.g., ``health,'' ``healthful,'' ``healthfully,'' ``healthfulness,'' ``healthier,'' ``healthiest,'' ``healthily,'' and ``healthiness'') as an implied nutrient content claim on the label or in labeling of a food that is useful in creating a diet that is consistent with dietary recommendations if:
  (i) The food meets the following conditions for fat, saturated fat, cholesterol, and other nutrients:

If the food is ...	The fat level must be ...	The saturated fat level must be ...	The cholesterol level must be ...	The food must contain ...
(A) A raw fruit or vegetable	Low fat as defined in Sec. 101.62(b)(2)	Low saturated fat as defined in Sec. 101.62(c)(2)	The disclosure level for cholesterol specified in Sec. 101.13(h) or less	N/A
(B) A single-ingredient or a mixture of frozen or canned fruits and vegetables \1\	Low fat as defined in Sec. 101.62(b)(2)	Low saturated fat as defined in Sec. 101.62(c)(2)	The disclosure level for cholesterol specified in Sec. 101.13(h) or less	N/A
(C) An enriched cereal-grain product that conforms to a standard of identity in part 136, 137 or 139 of this chapter	Low fat as defined in Sec. 101.62(b)(2)	Low saturated fat as defined in Sec. 101.62(c)(2)	The disclosure level for cholesterol specified in Sec. 101.13(h) or less	N/A

Chart VI.F.1(e)
(Continued)

If the food is ...	The fat level must be ...	The saturated fat level must be ...	The cholesterol level must be ...	The food must contain ...
(D) A raw single-ingredient seafood or game meat	Less than 5 grams (g) total fat per RA\2\ and per 100g	Less than 2 g saturated fat per RA and per 100 g	Less than 95 mg cholesterol per RA and per 100g	At least 10 percent of RDI\3\ or the DRV\4\ per RA of one or more of vitamin A, vitamin C, calcium, iron, protein, or fiber
(E) A meal product as defined in Sec. 101.13(h) or a main dish product as defined in Sec. 101.13(m)	Low fat as defined in Sec. 101.62(b)(3)	Low saturated fat as defined in Sec. 101.62(c)(3)	90 mg or less	At least 10 percent of RDI or DRV per LS of two nutrients (for a main dish product) or of three nutrients (for a meal product) of: vitamin A, vitamin C, calcium, iron, protein, or fiber
(F) A food not specifically listed in this table	Low fat as defined in Sec. 101.62(b)(2)	Low saturated fat as defined in Sec. 101.62(c)(2)	The disclosure level for cholesterol specified in Sec. 101.13(h) or less	At least 10 percent of the RDI or the DRV per RA of one or more of vitamin A, vitamin C, calcium, iron, protein, or fiber

\1\ May include ingredients whose addition does not change the nutrient profile of the fruit or vegetable.

\2\ RA means Reference Amount Customarily Consumed per Eating Occasion (Sec. 101.12(b)).

\3\ RDI means Reference Daily Intake (Sec. 101.9(c)(8)(iv)).

\4\ DRV means Daily Reference Value (Sec. 101.9(c)(9)).

\5\ LS means Labeled Serving, i.e., the serving size that is specified in the nutrition information on the product label (Sec. 101.9(b)).

Chart VI.F.1(e)
(Continued)

Chapter VI

(ii) The food meets the following conditions for sodium:	
If the food is...	**The sodium level must be...**
(A) A food with a RA that is greater than 30 g or 2 tablespoons (tbsp.)	480 mg or less sodium per RA per LS
(B) A food with a RA that is equal to or less than 30 g or 2 tbsp.	480 mg or less sodium per 50 g\1\
C) A meal product as defined in Sec. 101.13(1) or a main dish product as defined in Sec 101.13(m)	600 mg or less sodium per LS

\1\ For dehydrated food that is typically reconstituted with water or a liquid that contains insignificant amounts per RA of all nutrients (as defined in Sec. 101.9(f)(1)), the 50 g refers to the ``prepared''
form of the product.

(iii) The food complies with the definition and declaration requirements in this part 101 for any specific nutrient content claim on the label or in labeling, and
(iv) If you add a nutrient to the food specified in paragraphs (d)(2)(i)(D), (d)(2)(i)(E), or (d)(2)(i)(F) of this section to meet the 10 percent requirement, that addition must be in accordance with the fortification policy for foods in Sec. 104.20 of this chapter.

From 21 CFR Parts 100 through 169 (4/1/06 Edition)

(a) Summary Comments

Health claims describe a relationship between a food, food component, or dietary supplement ingredient, and reducing risk of a disease or health-related condition. There are three ways by which FDA exercises its oversight in determining which health claims may be used on a label or in labeling for a food or dietary supplement: 1) the 1990 Nutrition Labeling and Education Act (NLEA) provides for FDA to issue regulations authorizing health claims for foods and dietary supplements after FDA's careful review of the scientific evidence submitted in health claim petitions; 2) the 1997 Food and Drug Administration Modernization Act (FDAMA) provides for health claims based on an authoritative statement of a scientific body of the U.S. government or the National Academy of Sciences; such claims may be used after submission of a health claim notification to FDA; and 3) the 2003 FDA *Consumer Health Information for Better Nutrition Initiative* provides for qualified health claims where the quality and strength of the scientific evidence falls below that required for FDA to issue an authorizing regulation. Such health claims must be qualified to assure accuracy and non-misleading presentation to consumers.

A "health claim" by definition has two essential components: (1) a substance (whether a food, food component, or dietary ingredient) and (2) a disease or health-related condition. A statement lacking either one of these components does not meet the regulatory definition of a health claim. For example, statements that address a role of dietary patterns or of general categories of foods (e.g., fruits and vegetables) in health are considered to be dietary guidance rather than health claims, provided that the context of the statement does not suggest that a specific substance is the subject. Dietary guidance statements used on food labels must be truthful and non-misleading.

The FD&CA authorizes health claims, a claim characterizing the relationship between a substance and a disease or health-related condition, on food labels and labeling provided specific criteria are met, following the submission of a petition and promulgation of a health claim regulation. (section 403(r)(1)(B) of the FD&CA).

The Nutrition Labeling and Education Act (NLEA) of 1990, the Dietary Supplement Act of 1992, and the Dietary Supplement Health and Education Act of 1994 (DSHEA), provide for health claims used on labels that characterize a relationship between a food, a food component, dietary ingredient, or dietary supplement and risk of a disease (for example, "diets high in calcium may reduce the risk of osteoporosis"), provided the claims meet certain criteria and are authorized by an FDA regulation. FDA authorizes these types of health claims based on an extensive review of the scientific literature, generally as a result of the submission of a health claim petition, using the significant scientific agreement standard to determine that the nutrient/disease relationship is well established. (See Chart VI.F.2(c) Examples of Approved Health Claims)

(b) Questions and Responses

The questions and responses sections included under Chapter VI are a series of established requirements designed to aid the reviewers in establishing the degree of compliance that a specific food label and its labeling comply with applicable laws and regulations. It is not intended to represent all questions that may arise during the review process, but as examples of typical question during a label review. This section can also serve as a teaching aid for less experienced reviewers and as a ready reference for experienced reviewers. It can also serve as an example for developing a response to advise those responsible for food labels and labeling of needed corrections.

 Does the label or labeling of a food bear a claim, including written statements (e.g., a brand name including a term such as "heart"), symbols (e.g., a heart symbol), or vignettes, that expressly or by implication, including third party

references, characterizes the relationship of any substances to a disease or health-related condition? *(21 CFR 101.14(a)(1))*

A. If "*YES*," the food is subject to the health claims requirements under *21 CFR 101.14* and the Specific Requirements for Health Claims under *21 CFR Part 101, Subpart E*, continue.
B. If "*NO*," STOP HERE. The product is not subject to requirements for health claims.

Is the health claim presented on the label or in labeling authorized by regulation?

A. If "*YES*," continue.
B. If "*NO*," state that the label or labeling bears a health claim that is prohibited in that it has not been specifically provided for by regulation.

Does the food contain levels of fat, saturated fat, cholesterol, and sodium that fall below the disqualifying levels for these nutrients and thus is able to bear a health claim? *(See Chart VI.F.2(d))*

A. If "*YES*," continue.
B. If "*NO*," state that the product's label bears the health claim "_____" (insert the health claim as stated on the label or labeling), but the product exceeds the disqualifying nutrient levels of _____ (insert the nutrient(s) and level(s) for the food) for the claim.

Is the health claim complete, truthful, and not misleading as required by the regulations in 21 CFR Part 101, Subpart E?

A. If "*YES*," continue.
B. If "*NO*," state how the claim is false, incomplete, and/or misleading.

Does the label or labeling of a food bear a claim, including written statements (e.g., a brand name including a term such as "heart"), symbols (e.g., a heart symbol), or vignettes, that expressly or by implication, including third party references, characterizes the relationship of any substances to a disease or health-related condition? *(21 CFR 101.14(a)(1))*

A. If "*YES*," the food is subject to the health claims requirements under *21 CFR 101.14* and the Specific Requirements for Health Claims under *21 CFR Part 101, Subpart E*, continue.
B. If "*NO*," STOP HERE. The product is not subject to requirements for health claims.

Is the health claim presented on the label or in labeling authorized by regulation?

A. If "*YES*," continue.
B. If "*NO*," state that the label or labeling bears a health claim that is prohibited in that it has not been specifically provided for by regulation.

Does the label or labeling of a food bear a claim, including written statements (e.g., a brand name including a term such as "heart"), symbols (e.g., a heart symbol), or vignettes, that expressly or by implication, including third party references, characterizes the relationship of any substances to a disease or health-related condition? *(21 CFR 101.14(a)(1))*

A. If "*YES*," the food is subject to the health claims requirements under *21 CFR 101.14* and the Specific Requirements for Health Claims under *21 CFR Part 101, Subpart E*, continue.

B. If *"NO,"* STOP HERE. The product is not subject to requirements for health claims.

 Q ***Is the health claim presented on the label or in labeling authorized by regulation?***

A. If *"YES,"* continue.
B. If *"NO,"* state that the label or labeling bears a health claim that is prohibited in that it has not been specifically provided for by regulation.

Approved Claims	Food Requirements	Claim Requirements	Model Claim, Statements
Calcium and Osteoporosis-- *21 CFR 101.72*	- High in calcium, - Assimilable (Bioavailable), - Supplements must disintegrate and dissolve, and - Phosphorus content cannot exceed calcium content	Indicates disease depends on many factors by listing risk factors or the disease: Gender--Female. Race--Caucasian and Asian. Age--Growing older. Primary target population: Females, Caucasian and Asian races, and teens and young adults in their bone-forming years. Additional factors necessary to reduce risk: Eating healthful meals, regular exercise. Mechanism relating calcium to osteoporosis: Optimizes peak bone mass. Foods or supplements containing more than 400 mg calcium must state that total intakes of greater than 2,000 mg calcium provide no added benefit to bone health.	Regular exercise and a healthy diet with enough calcium helps teens and young adult white and Asian women maintain good bone health and may reduce their high risk of osteoporosis later in life.
Sodium and Hypertension-- *21 CFR 101.74*	- Low sodium	*Required terms:* - "Sodium", "High blood pressure" Includes physician statement (Individuals with high blood pressure should consult their physicians) if claim defines high or normal blood pressure.	Diets low in sodium may reduce the risk of high blood pressure, a disease associated with many factors.
Dietary Lipids (Fat) and Cancer-- *21 CFR 101.73*	- Low fat (Fish & game meats: "Extra lean")	*Required terms:* - "Total fat" or "Fat" - "Some types of cancers" or "Some cancers"	Development of cancer depends on many factors. A diet low in total fat may reduce the risk of some cancers.

 Adopted from FDA, CFSAN, Food Labeling Guide, September 1994 (Editorial Revisions June 1999 and November 2002)

Chart VI.F.2(c)

(Continued)

Approved Claims	Food Requirements	Claim Requirements	Model Claim, Statements
		Does not specify types of fats or fatty acids that may be related to risk of cancer.	
Dietary Saturated Fat and Cholesterol and Risk of Coronary Heart Disease-- *21 CFR 101.75*	- Low saturated fat, - Low cholesterol, and - Low fat (Fish & game meats: "Extra lean")	*Required terms:* - "Saturated fat and cholesterol", - "Coronary heart disease" or "Heart disease" Includes physician statement (individuals with elevated blood total--or LDL--cholesterol should consult their physicians) if claim defines high or normal blood total--and LDL--cholesterol.	While many factors affect heart disease, diets low in saturated fat and cholesterol may reduce the risk of this disease.
Fiber-Containing Grain Products, Fruits, and Vegetables and Cancer-- *21 CFR 101.76*	- A grain product, fruit, or vegetable that contains dietary fiber; - Low fat, and - Good source of dietary fiber (without fortification)	*Required terms:* - "Fiber", "Dietary fiber", or "Total dietary fiber" - "Some types of cancer" or "Some cancers" Does not specify types of dietary fiber that may be related to risk of cancer.	Low fat diets rich in fiber-containing grain products, fruits, and vegetables may reduce the risk of some types of cancer, a disease associated with many factors.
Fruits, Vegetables and Grain Products that contain Fiber, particularly Soluble Fiber, and Risk of Coronary Heart Disease- - *21 CFR 101.77*	- A fruit, vegetable, or grain product that contains fiber; - Low saturated fat, - Low cholesterol, - Low fat, - At least 0.6 grams of soluble fiber per RA (without fortification), and, - Soluble fiber content provided on label	*Required terms:* - "Fiber", "Dietary fiber", "Some types of dietary fiber", "Some dietary fibers", or "Some fibers" - "Saturated fat" and "Cholesterol" - "Heart disease" or "Coronary heart disease" Includes physician statement ("Individuals with	Diets low in saturated fat and cholesterol and rich in fruits, vegetables, and grain products that contain some types of dietary fiber, particularly soluble fiber, may reduce the risk of heart disease, a disease associated with many factors.

Chart VI.F.2(c)
(Continued)

Chapter VI

Approved Claims	Food Requirements	Claim Requirements	Model Claim, Statements
		elevated blood total-- or LDL--cholesterol should consult their physicians") if claim defines high or normal blood total--and LDL-- cholesterol.	
Fruits and Vegetables and Cancer-- *21 CFR 101.78*	- A fruit or vegetable, - Low fat, and - Good source (without fortification) of at least one of the following: • Vitamin A, • Vitamin C, or • Dietary fiber	*Required terms:* - "Fiber", "Dietary fiber", or "Total dietary fiber"; - "Total fat" or "Fat", - "Some types of cancer" or "Some cancers" Characterizes fruits and vegetables as "Foods that are low in fat and may contain Vitamin A, Vitamin C, and dietary fiber." Characterizes specific food as a "Good source" of one or more of the following: Dietary fiber, Vitamin A, or Vitamin C. Does not specify types of fats or fatty acids or types of dietary fiber that may be related to risk of cancer.	Low fat diets rich in fruits and vegetables (foods that are low in fat and may contain dietary fiber, Vitamin A, or Vitamin C) may reduce the risk of some types of cancer, a disease associated with many factors. Broccoli is high in vitamin A and C, and it is a good source of dietary fiber.
Folate and Neural Tube Defects—*21 CFR 101.79*	"Good Source" of folate (at least 40 mcg folate per serving) -Dietary supplements, or foods in conventional food form that are naturally good sources of folate (i.e., only non-fortified food in conventional food form) -The claim shall not be made on products that contain more than 100% of the vitamin A as retinol or performed	*Required terms:* -Terms that specify the relationship (e.g., women who are capable of becoming pregnant and who consume adequate amounts of folate) "Folate", "folic acid", "folacin", "folate a B vitamin", "folic acid, a B vitamin," "folacin, a B vitamin," "neural tube defects", "birth defects, spinal bifida, or anencephaly", "birth defects of the	Healthful diets with adequate folate may reduce a woman's risk of having a child with brain or spinal cord defect.

Adopted from FDA, CFSAN, Food Labeling Guide, September 1994 (Editorial Revisions June 1999 and November 2002)

Chart VI.F.2(c)
(Continued)

Approved Claims	Food Requirements	Claim Requirements	Model Claim, Statements
	vitamin A or vitamin D Dietary supplements shall meet USP standards for disintegration and dissolution or otherwise bioavailable	brain or spinal cord—anencephaly, birth defects of the brain or spinal cord" Must also include information on the multifactorial nature of neural tuber defects, and the safe upper limit of daily intake.	
Dietary Sugar Alcohol and Dental Caries-- *21 CFR 101.80*	-Sugar free -The sugar alcohol must be xylitol, sorbitol, mannitol, maltitol, isomalt, lactitol, hydrogenated starch hydrolysates, hydrogenated glucose syrups, erythritol, or a combination -When a fermentable carbohydrate is present, the food must not lower plaque pH below 5.7	*Required terms:* -"does not promote," "may reduce the risk of," "useful [or is useful] in not promoting" or "expressly [or is expressly] for not promoting" dental caries; -"sugar alcohol" or "sugar alcohols" or the name or names of the sugar alcohols, e.g., sorbitol; -"dental caries" or "tooth decay" Includes statement that frequent between meal consumption of foods high in sugars and starches can promote tooth decay. Packages with less than 15 square inches of surface area available for labeling may use a shortened claim.	**Full claim**: Frequent between-meal consumption of foods high in sugars and starches promotes tooth decay. The sugar alcohols in [name of food] do not promote tooth decay. **Shortened claim (on small packages only)**: Does not promote tooth decay.
Soluble Fiber from Certain Foods and Risk of Coronary Heart Disease - *21 CFR 101.81*	-Low saturated fat -Low cholesterol -Low fat -Include either (1) one or more eligible sources of whole oats, containing at	*Required terms:* - "Heart disease" or "coronary heart disease." - "Soluble fiber" qualified by either "psyllium seed husk" or the name of the	Soluble fiber from foods such as [*name of soluble fiber source, and, if desired, name of food product*], as part of a diet low in saturated fat and cholesterol, may reduce the risk of heart disease. A serving of [*name of food product*]

Chart VI.F.2(c)
(Continued)

Chapter VI

Approved Claims	Food Requirements	Claim Requirements	Model Claim, Statements
	least 0.75 g whole oat soluble fiber per RA; or (2) psyllium seed husk containing at least 1.7 g of psyllium husk soluble fiber per RA -Amount of soluble fiber per RA declared in nutrition label. **Eligible Source of Soluble Fiber** Beta glucan soluble fiber from oat bran, rolled oats (or oatmeal), and whole oat flour. Oat bran must provide at least 5.5% beta-glucan soluble fiber, rolled oats must provide at least 4% beta-glucan soluble fiber, and whole oat flour must provide at least 4% beta-glucan soluble fiber or Psyllium husk with purity of no less than 95% Whole grain barley and dry milled products provide at least 3 grams of beta-glucan soluble fiber per day	eligible source of whole oat soluble fiber. - "Saturated fat" and "cholesterol." - "Daily dietary intake of the soluble fiber source necessary to reduce the risk of CHD and the contribution one serving of the product makes to this level of intake." **Additional Required Label Statement** Foods bearing a psyllium seed husk health claim must also bear a label statement concerning the need to consume them with adequate amounts of fluids; e.g., "NOTICE: This food should be eaten with at least a full glass of liquid. Eating this product without enough liquid may cause choking. Do not eat this product if you have difficulty in swallowing." (*21 CFR 101.17(f)*)	supplies __ grams of the [necessary daily dietary intake for the benefit] soluble fiber from [*name of soluble fiber source*] necessary per day to have this effect. The health claim has been broadened to include whole grain barley and dry milled products as additional sources of beta-glucan soluble fiber eligible for the health claim.

Chart VI.F.2(c)
(Continued)

Approved Claims	Food Requirements	Claim Requirements	Model Claim, Statements
Soy Protein and Risk of Coronary Heart Disease *21 CFR 101.82*	- At least 6.25 g soy protein per RA - Low saturated fat, - Low cholesterol, and - Low fat (except that foods made from whole soybeans that contain no fat in addition to that inherent in the whole soybean are exempt from the "low fat" requirement)	*Required terms:* - "Heart disease" or "coronary heart disease" - "Soy protein" - "Saturated fat" and "cholesterol" Claim specifies daily dietary intake levels of soy protein associated with reduced risk. Claim specifies amount of soy protein in a serving of food.	(1) 25 grams of soy protein a day, as part of a diet low in saturated fat and cholesterol, may reduce the risk of heart disease. A serving of [*name of food*] supplies ___ grams of soy protein. (2) Diets low in saturated fat and cholesterol that include 25 grams of soy protein a day may reduce the risk of heart disease. One serving of [*name of food*] provides ___ grams of soy protein.
Plant Sterol/stanol esters and Risk of Coronary Heart Disease *21 CFR 101.83*	- At least 0.65 g plant sterol esters per RA of spreads and salad dressings, or - At least 1.7 g plant stanol esters per RA of spreads, salad dressings, snack bars, and dietary supplements. - Low saturated fat, - Low cholesterol, and - Spreads and salad dressings that exceed 13 g fat per 50 g must bear the statement "*see nutrition information for fat content*" Salad dressings are exempted from the minimum 10% DV nutrient requirement (see General Criteria below)	*Required terms:* - " May" or " might" reduce the risk of CHD - " Heart disease" or " coronary heart disease" - "Plant sterol esters" or " plant stanol esters"; except "vegetable oil" may replace the term "plant" if vegetable oil is the sole source of the sterol/stanol ester Claim specifies plant stero/stanol esters are part of a diet low in saturated fat and cholesterol. Claim does not attribute any degree of CHD risk reduction. Claim specifies the daily dietary intake of plant sterol or stanol esters necessary to reduce CHD risk, and the amount provided per serving.	(1) Foods containing at least 0.65 gram per serving of vegetable oil sterol esters, eaten twice a day with meals for a daily total intake of at least 1.3 grams, as part of a diet low in saturated fat and cholesterol, may reduce the risk of heart disease. A serving of [*name of food*] supplies ___ grams of vegetable oil sterol esters. (2) Diets low in saturated fat and cholesterol that include two servings of foods that provide a daily total of at least 3.4 grams of plant stanol esters in two meals may reduce the risk of heart disease. A serving of [*name of food*] supplies ___ grams of plant stanol esters.

Adopted from FDA, CFSAN, Food Labeling Guide, September 1994 (Editorial Revisions June 1999 and November 2002, and revised 5/22/06)

155

Chart VI.F.2(c)
(Continued)

Chapter VI

Approved Claims	Food Requirements	Claim Requirements	Model Claim, Statements
		Claim specifies that plant sterol or stanol esters should be consumed with two different meals each a day.	

Chart VI.F.2(d)
Health Claims Criteria and
Disqualifying Nutrient Levels

General Criteria All Claims Must Meet

- All information in one place without intervening material (Reference statement permitted).

- Only information on the value that intake or reduced intake, as part of a total dietary pattern, may have on a disease or health-related condition.

- Enables public to understand information provided and significance of information in the context of a total daily diet.

- Complete, truthful, and not misleading.

- Food contains, without fortification, 10% or more of the Daily Value for one of six nutrients (dietary supplements excepted):

Vitamin A	500 IU	Calcium	100 mg
Vitamin C	6 mg	Protein	5 g
Iron	1.8 mg	Fiber	2.5 g

- Not represented for infants or toddlers less than 2 years of age.

- Uses "may" or "might" to express relationship between substance and disease.

- Does not quantify any degree of risk reduction.

- Indicates disease depends on many factors.

- Food contains less than the specified levels of four disqualifying nutrients:

Disqualifying Nutrients	Foods	Main Dishes	Meal Products
Fat	13 g	19.5 g	26 g
Saturated Fat	4 g	6 g	8 g
Cholesterol	60 mg	90 mg	120 mg
Sodium	480 mg	720 mg	960 mg

Abbreviations: RA = reference amount, IU = International Units

Approved Health Claims

- **Calcium and Osteoporosis**
    - *21 CFR 101.72* Health claims: calcium and osteoporosis.

- **Dietary Lipids (Fat) and Cancer**
    - *21 CFR 101.73* Health claims: dietary lipids and cancer.

- **Dietary Saturated Fat and Cholesterol and Risk of Coronary Heart Disease**
    - *21 CFR 101.75* Health claims: dietary saturated fat and cholesterol and risk of coronary heart disease.

- **Dietary Non-cariogenic Carbohydrate Sweeteners and Dental Caries**
    - *21 CFR 101.80* Health claims: dietary sugar alcohols and dental caries.
    - Final Rule: Health Claims: D-tagatose and Dental Caries. July 3, 2003
    - Interim Final Rule: Health Claims: D-tagatose and Dental Caries. December 2, 2002
    - Final Rule: Food Labeling: Health Claims: Dietary Sugar Alcohols and Dental Caries. December 2, 1997
    - Final Rule: Food Labeling: Health Claims: Sugar Alcohol and Dental Caries. August 23, 1996

- **Fiber-containing Grain Products, Fruits and Vegetables and Cancer**
    - *21 CFR 101.76* Health claims: fiber-containing grain products, fruits, and vegetables and cancer.

- **Folic Acid and Neural Tube Defects**
    - *21 CFR 101.79* Health claims: Folate and neural tube defects.
    - Final Rule: Food Labeling: Health Claims: Folate and Neural Tube Defects. March 5, 1996
    - Final Rule: Revoking January 4, 1994 Regulation That Became Final By Operation of Law. September 24, 1996

- **Fruits and Vegetables and Cancer**
    - *21 CFR 101.78* Health claims: fruits and vegetables and cancer.

- **Fruits, Vegetables and Grain Products that contain Fiber, particularly Soluble Fiber, and Risk of Coronary Heart Disease**
    - *21 CFR 101.77* Health claims: fruits, vegetables, and grain products that contain fiber, particularly soluble fiber, and risk of coronary heart disease.

- **Soluble Fiber from Certain Foods and Risk of Coronary Heart Disease**
    - *21 CFR 101.81* Health claims: Soluble fiber from certain foods and risk of coronary heart disease (CHD).
    - Final Rule: Health Claims: Soluble Dietary Fiber from certain foods and risk of coronary heart disease (CHD) (Barley). May 22, 2006
    - Interim final rule: Health Claims: Soluble Dietary Fiber From Certain Foods and Coronary Heart Disease (Barley) December 23, 2005
    - Final Rule: Health Claims: Soluble Dietary Fiber From Certain Foods and Coronary Heart Disease (Oatrim) July 28, 2003
    - Interim Final Rule: Health Claims: Soluble Dietary Fiber From Certain Foods and Coronary Heart Disease (Oatrim) October 2, 2002
    - Final Rule; correction: Food Labeling: Health Claims: Soluble Fiber From Certain Foods and Coronary Heart Disease: Correction (Psyllium husk). April 9, 1998

Chart VI.F.2(e)
(Continued)

- o Final Rule: <u>Food Labeling: Health Claims: Soluble Fiber from Certain Foods and Risk of Coronary Heart Disease (Psyllium husk).</u> February 18, 1998
- o Final Rule: <u>Food Labeling: Health Claims: Soluble Fiber From Whole Oats and Risk of Coronary Heart Disease.</u> Amended. March 31, 1997
- o Final Rule: <u>Food Labeling: Health Claims: Oats and Coronary Heart Disease.</u> January 23,1997
- o <u>FDA Allows Whole Oat Foods to Make Health Claim on Reducing the Risk of Heart Disease.</u> January 21, 1997

- **Soy Protein and Risk of Coronary Heart Disease**
  - o *21 CFR 101.82* <u>Health claims: Soy Protein and risk of coronary heart disease (CHD).</u>
  - o <u>FDA Approves New Health Claim for Soy Protein and Coronary Heart Disease.</u> October 20, 1999
  - o Final Rule: <u>Food Labeling: Health Claims: Soy Protein and Coronary Heart Disease.</u> October 26, 1999
  - o <u>Soy: Health Claims for Soy Protein. Questions About Other Components.</u> May-June 2000
  - o <u>Other Information on Soy</u>

- **Stanols/Sterols and Risk of Coronary Heart Disease**
  - o <u>FDA Letter Regarding: Enforcement Discretion With Respect to Expanded Use of an Interim Health Claim Rule About Plant</u>
  - o <u>Sterol/Stanol Esters and Reduced Risk of Coronary Heart Disease.</u> February 14, 2003
  - o *21 CFR 101.83* <u>Health claims: plant sterol/stanol esters and risk of coronary heart disease (CHD).</u>
  - o <u>FDA Authorizes New Coronary Heart Disease Claim for Plant Sterol and Plant Stanol Esters.</u> September 5, 2000
  - o <u>Interim Final Rule - Food Labeling: Health Claims: Plant Sterol/Stanol Esters and Coronary Heart Disease.</u> September 8, 2000
  - o <u>Interim Final Rule: Reopening of Comment Period – Food Labeling: Health Claims: Plant Sterol/Stanol Esters and Coronary Heart Disease.</u> October 5, 2001

(a) Summary Comments

Prior to the Food and Drug Administration Modernization Act of 1997 (FDAMA), companies could not use a health claim or nutrient content claim in food labeling unless the FDA published a regulation authorizing such a claim. Two new provisions of FDAMA (specifically sections 303 and 304 which amend, respectively, sections 403(r)(3) and 403(r)(2) of the FD&CA) will now permit distributors and manufacturers to use claims if such claims are based on current, published, authoritative statements from certain federal scientific bodies, as well as from the National Academy of Sciences. These provisions are intended to expedite the process by which the scientific basis for such claims is established.

FDAMA permits claims based on current, published authoritative statements from "a scientific body of the United States with official responsibility for public health protection or research directly related to human nutrition... or the National Academy of Sciences (NAS) or any of its subdivisions." The National Institutes of Health (NIH) and the Centers for Disease Control and Prevention (CDC) are federal government agencies specifically identified as scientific bodies by FDAMA.

FDA stated that other federal agencies might also qualify as appropriate sources for such authoritative statements. Along with NAS (or any of its subdivisions), the agency currently considers that the following federal scientific bodies may be sources of authoritative statements: the CDC, the NIH, and the Surgeon General within Department of Health and Human Services; and the Food and Nutrition Service, the Food Safety and Inspection Service, and the Agriculture Research Service within the Department of Agriculture.

FDAMA states that an authoritative statement: (1) is "about the relationship between a nutrient and a disease or health-related condition" for a health claim, or "identifies the nutrient level to which the claim refers" for a nutrient content claim, (2) is "published by the scientific body" (as identified above), (3) is "currently in effect," and (4) "shall not include a statement of an employee of the scientific body made in the individual capacity of the employee." In addition, FDA has stated that authoritative statements also should (5) reflect a consensus within the identified scientific body if published by a subdivision of one of the Federal scientific bodies, and (6) be based on a deliberative review by the scientific body of the scientific evidence.

Not all pronouncements by the designated scientific bodies would meet these criteria. For example, authoritative statements by the Surgeon General would normally be found only in the Surgeon General Reports.

FDA intends to consult, as appropriate, with the scientific body that is the source of a statement cited as the basis for a claim, as well as with the other federal scientific bodies that have public health responsibilities and expertise relative to the claim.

For information and guidance purposes on authoritative statements, this manual includes: (1) the regulation in *21 CFR 101.90* "Notifications for Health Claims Based on Authoritative Statements" (See Regulations VI.F.3(b)); (2) "Interim Final Rules Prohibiting Health Claims Based on Authoritative Statements" (See Chart VI.F.3(c)); and (3) "Allowable Authoritative Statements by Action of the Statute" (See Chart VI.F.3(d)).

Regulation VI.F.3(b)
Notifications of Health Claims Based on
"Authoritative Statements"
(Proposed 21 CFR 101.90)

a) A claim of the type described in *§ 101.14(a)(1)* which is not authorized by the Food and Drug Administration (FDA) in a regulation found in this part shall be authorized and may be made with respect to a dietary supplement if:

   (1) A scientific body of the U.S. Government with official responsibility for public health protection or research directly relating to human nutrition (such as the National Institutes of Health or the Centers for Disease Control and Prevention) or the National Academy of Sciences or any of its subdivisions has published an authoritative statement, which is currently in effect, about the relationship between a nutrient and a disease or health-related condition to which the claim refers;

   (2) A person has submitted to FDA, at least 120 days (during which FDA may notify any person who is making a claim as authorized by paragraph (a) of this section that such person has not submitted all the information required by this paragraph) before the first introduction into interstate commerce of the dietary supplement with a label containing the claim:

      (i) A notice of the claim, which shall include the exact words used in the claim and shall include a concise description of the basis upon which such person relied for determining that the requirements of paragraph (a)(1) of this section have been satisfied;

      (ii) A copy of the statement referred to in paragraph (a)(1) of this section upon which such person relied in making the claim; and

      (iii) A balanced representation of the scientific literature relating to the relationship between a nutrient and a disease or health-related condition to which the claim refers;

   (3) The claim and the dietary supplement for which the claim is made are in compliance with *§ 101.14(a)(5)* and *(e)(3)* and are otherwise in compliance with sections 403(a) and 201(n) of the act (21 U.S.C. 343(a) and 21 U.S.C. 321(n)); and

   (4) The claim is stated in a manner so that the claim is an accurate representation of the authoritative statement referred to in paragraph (a)(1) of this section and so that the claim enables the public to comprehend the information provided in the claim and to understand the relative significance of such information in the context of a total daily diet. For purposes of this paragraph, a statement shall be regarded as an authoritative statement of a scientific body described in paragraph (a)(1) of this section only if the statement is published by the scientific body and shall not include a statement of an employee of the scientific body made in the individual capacity of the employee.

b) A claim submitted under the requirements of paragraph (a) of this section may be made until:

   (1) Such time as FDA issues a regulation under the standard in *§ 101.14(c)*:

      (i) Prohibiting or modifying the claim and the regulation has become effective; or

      (ii) Finding that the requirements of paragraph (a) of this section have not been met and has not submitted all the information required by such clause; or

   (2) A District Court of the United States has ruled, including finding that the petitioner enforcement proceeding under chapter III of the act (21 U.S.C. 301–310) has determined that the requirements of paragraph (a) of this section have not been met.

Claim	Statement	Authoritative Body	Basis for Denial
Health Claim: Antioxidant vitamins C and E may reduce the risk in adults of atherosclerosis, coronary heart disease, certain cancers, and cataracts. FR 63, 6/22/99, 34083-34091	1. "Antioxidant micronutrients, especially carotenes, vitamin C, and vitamin E, appear to play many important roles in protecting the body against cancer. They block the formation of chemical carcinogens in the stomach, protect DNA and lipid membranes from oxidative damage, and enhance immune function."	1. A published article authorized by two employees of Centers for Disease Control and Prevention (CDC).	CDC stated that the statement is not an "authoritative statement" of CDC because it does not reflect consensus within CDC and was not published by CDC.
	2. "[Antioxidants] may help prevent disease. Antioxidants fight harmful molecules called oxygen free radicals, which are created by the body as cells go about their normal business of producing energy... [some] studies show that antioxidants may help prevent heart disease, some cancers, and cataracts that are more common as people get older."	2. Public information provided on the Internet by an Institute of NIH.	NIH responded that the statement is not an authoritative statement of NIH because it is not based on a deliberative review of the scientific evidence regarding the nutrient-disease relationship in question.
	3. "The antioxidant nutrients found in plant foods (e.g., vitamin C, carotenoids, vitamin E, and certain minerals) are presently of great interest to scientists and the public because of their potentially beneficial role in reducing the risk of cancer and certain other chronic diseases."	3. An electronic version provided on the Internet on "Nutrition and Your Health: Dietary Guidelines for Americans" (recommendations developed by a group of Federal agencies and issued jointly by the DHHS and USDA).	FDA has authorized a health claim for the relationship between cancer and fruits and vegetables that contain vitamin C, as well as vitamin A (beta-carotene), and dietary fiber, under *21 CFR 101.78*. On this basis, the statement is not an "authoritative statement" under section 403(r)(3)(C) of the Act because the statement indicates that the scientific evidence about the

Adopted from the Federal Register (63 FR 34084-34115; (June 22, 1998)

Claim	Statement	Authoritative Body	Basis for Denial
Health Claim: Antioxidant vitamins C and E may reduce the risk in adults of atherosclerosis, coronary heart disease, certain cancers, and cataracts. FR 63, 6/22/99, 34083-34091 (Continued)	4. "A diet high in fiber, high in antioxidants, and low in fat may play an important role in preventing the development of atherosclerosis, coronary heart disease, and some cancers."	4. Public information provided on the Internet by CDC's Office of Women's Health.	relationship in question is preliminary or inconclusive. CDC stated that the statement is not an authoritative statement of CDC because, although it is a statement from CDC, it is not based upon a deliberative review of the scientific evidence regarding the nutrient-disease relationship in question.
	5. "[It] is likely that certain antioxidants, such as vitamins C and E, may destroy the oxygen radicals, retard molecular damage, and perhaps slow the rate of aging."	5. An NIH press release provided on the Internet.	The statement does not address a disease or health-related condition and, therefore, is not an "authoritative statement" under section 403(r)(3)(C) of the Act.
	6. "Antioxidants are thought to help prevent heart attack, stroke, and cancer."	6. An electronic version provided on the Internet of a quarterly report from USDA's ARS.	USDA stated that the statement is not an "authoritative statement" of USDA because it was not based upon a deliberative review of the scientific evidence regarding a relationship between the nutrient and the disease in question.
Health Claim: Vitamin K and promotion of proper blood clotting and improvement in bone health in adults. FR 63, 6/22/98, 34115-34117	"In adults, vitamin K promotes proper blood clotting and may improve bone health. Sources of vitamin K include spinach, cabbage, turnip greens, broccoli, tomatoes, and dietary supplements."		The proposed claim is prohibited as a health claim because it does not characterize the relationship of the nutrient vitamin K to a disease or health-related condition.

Chart VI.F.3(c)
(Continued)

Claim	Statement	Authoritative Body	Basis for Denial
Health Claim: Garlic, reduction of serum cholesterol, and the risk of cardiovascular disease in adults. FR 63, 6/22/98, 34110-34112	"In adults, garlic may reduce serum cholesterol and the risk of cardiovascular disease."		The proposed claim is prohibited as a health claim because the statement submitted as the basis of the claim is not an "authoritative statement" of a scientific body.
Health Claim: Omega-3 fatty acids and the risk in adults of cardiovascular disease. FR 63, 6/22/98, 34107-34110	1. "Intake of particular polyunsaturated fats, the omega-3 fatty acids, may offer some protection against the development of clinical manifestations of atherosclerosis by decreasing platelets aggregation and clotting activity and preventing arterial thrombosis."	1. Contained in "Nutrition Monitoring in the United States—An Update Report on Nutrition Monitoring," prepared for USDA and the Public Health Service of DHHS by LSRO of the FASEB.	The proposed claim is prohibited because the statements submitted as the basis of the claim are not "authoritative statements" of a scientific body. The report was prepared under contract by LSRO/FASEB, an organization that is neither a Federal Government agency nor affiliated with National Academy of Sciences.
	2. "In new soybean oil varieties developed by USDA's ARS palmitic acid is replaced with oleic acid, which has some health benefits. In addition, omega-3 and omega-6 fatty acids, which can actually lower cholesterol levels, are at 7 and 60 percent respectively—essentially the same as regular soybeans."	2. Contained in a press release from USDA's ARS, entitled "New Soybeans Halve Saturated Fat, Keep Nutrition," which was provided on the Internet.	USDA stated that the statement is not an "authoritative statement" of USDA because it was not based upon a deliberative review of the scientific evidence regarding a relationship between the nutrient and the disease in question.
Health Claim: Zinc and the body's ability to fight infection and heal wounds in adults. FR 63, 6/22/98, 34112-34115	1. Zinc is an essential mineral in the diet and is a component of many enzymes. As such, it is involved in many metabolic processes including wound healing, immune function, growth, and maintenance of tissues."	1. Contained in "Nutrition Monitoring in the United States—An Update Report on Nutrition Monitoring," prepared for USDA and the Public Health Service of DHHS by LSRO of the FASEB.	The statement is not an "authoritative statement" because it does not reflect the official policy of an appropriate scientific body, nor has an appropriate scientific body conducted a deliberative review of the scientific evidence.

Adopted from the Federal Register (63 FR 34084-34115; (June 22, 1998)

Chart VI.F.3(c)
(Continued)

Claim	Statement	Authoritative Body	Basis for Denial
	2. "Dietary zinc shortages– a bigger problem in developing countries than in the United States–may be linked to depressed growth in children, slower wound-healing and difficult births."	2. Found in Human Nutrition (quarterly reports of selected research projects) issued by USDA's ARS and provided on the Internet.	USDA stated that the statement is not an authoritative statement of USDA because it was not based upon a deliberative review of the scientific evidence regarding a relationship between the nutrient and the disease in question.
Health Claim: Chromium and the risk in adults of hyperglycemia and the effects of glucose intolerance. FR 63, 6/22/98, 34104-34107	1. "Chromium supplements—in two different formulations—lowered blood pressure in rats bred to spontaneously develop hypertension … the supplements, chromium picolinate and chromium nicotinate, also reduced the formation of damaging free radicals in the animals' tissues, indicating that chromium can act as an antioxidant … chromium is essential for insulin to operate efficiently and has been shown to reduce diabetic symptoms and restore glucose tolerance in studies of humans and animals."	1. Found in Human Nutrition (quarterly reports of selected research projects) issued by USDA's ARS and provided on the Internet.	USDA stated that the statement is not an "authoritative statement" of USDA because it was not based upon a deliberative review of the scientific evidence regarding a relationship between the nutrient and the disease in question.
	2. "In a 20-week ARS study, rats that daily consumed more than 2,000 times the estimated safe limit of chromium for people showed no sign of toxicity … [the findings] brings into question the relevance of a study done 2 years ago … that reported DNA damage."	2. Found in Human Nutrition (quarterly reports of selected research projects) issued by USDA's ARS and provided on the Internet.	FDA concluded that the statement focuses on the levels of intake considered safe in rats and does not identify a relationship between a nutrient and disease or health-related condition in humans.

Chart VI.F.3(c)
(Continued)

Chapter VI

Claim	Statement	Authoritative Body	Basis for Denial
Health Claim: Chromium and the risk in adults of hyperglycemia and the effects of glucose intolerance. FR 63, 6/22/98, 34104-34107 (Continued)	3. "Scientists must often draw inferences about the relationships between dietary factors and disease from animal studies or human metabolic and population studies that approach issues indirectly."	3. Found in a discussion on the nature of scientific evidence contained in "The Surgeon General's Report on Nutrition and Health—Summary and Recommendations" that was published by the Public Health Service.	FDA concluded that the statement focuses on a general principal of scientific inference and is not about the relationship between a nutrient and a disease or health-related condition.
Health Claim: Calcium consumption by adolescents and adults, bone density and the risk of fractures. FR 63, 6/22/98, 34101-34104	1. "Although the precise relationship of dietary calcium to osteoporosis has not been elucidated, it appears that higher intakes of dietary calcium could increase peak bone mass during adolescence and delay onset of bone fractures later in life."	1. Excerpted from the Summary and Recommendations section of the 1988 Surgeon General's Report on Nutrition and Health.	These statements are about calcium and osteoporosis and are, therefore, prohibited because section 303 of FDAMA does not apply when the FDA has an existing regulation authorizing a health claim about the relationship between the nutrient and the disease or health-related condition at issue.
	2. "Inadequate dietary calcium consumption in the first three to four decades of life may be associated with increased risk of osteoporosis in later life."	2. Excerpted from the Summary and Recommendations section of the 1988 Surgeon General's Report on Nutrition and Health.	
	3. "[e]vidence shows that chronically low calcium intake, especially during adolescence and early adulthood, may compromise development of peak bone mass."	3. Excerpted from the Summary and Recommendations section of the 1988 Surgeon General's Report on Nutrition and Health.	

Adopted from the Federal Register (63 FR 34084-34115; (June 22, 1998)

Chart VI.F.3(c)
(Continued)

Claim	Statement	Authoritative Body	Basis for Denial
Health Claim: B-complex vitamins, lowered homocysteine levels, and the risk in adults of cardiovascular disease. FR 63, 6/22/98, 34097-34101	1. "A research team's new evidence confirms earlier data that elevated levels of the amino acid homocysteine increase the odds for significant narrowing of the arteries... The Analysis also Showed that Insufficient Levels of Folate and, to a Lesser Extent, Vitamin B_6 contribute to increased risk of artery narrowing. Like a see-saw, homocysteine levels go up as the vitamins go down, and vice versa."	1. Found in Human Nutrition (quarterly reports of selected research projects) and provided on the Internet. This paragraph is attributed to the USDA Human Nutrition Research Center on Aging at Tufts.	USDA stated that the statement is not an "authoritative statement" of USDA because it was not based upon a deliberative review of the scientific evidence regarding a relationship between the nutrient and the disease in question.
	2. "When people don't have enough of these [vitamin B_{12} and folate] vitamins to metabolize homocysteine it accumulates in the blood and damages the vessels."	2. Found in Human Nutrition (quarterly reports of selected research projects) which is issued by USDA and provided on the Internet.	USDA stated that the statement is not an "authoritative statement" of USDA because it was not based upon a deliberative review of the scientific evidence regarding a relationship between the nutrient and the disease in question.
	3. "[T]he body needs [folate] to convert homocysteine into a nontoxic amino acid and thus prevent damage to blood vessels ... Supplement users had the lowest homocysteine levels but not much lower than frequent consumers of fruits, vegetables and cereal."	3. Found in Human Nutrition (quarterly reports of selected research projects) which is issued by USDA and provided on the Internet.	USDA stated that the statement is not an "authoritative statement" of USDA because it was not based upon a deliberative review of the scientific evidence regarding a relationship between the nutrient and the disease in question.

Chart VI.F.3(c)	Chapter VI
(Continued)	

Claim	Statement	Authoritative Body	Basis for Denial
Health Claim: Antioxidant vitamin A and beta-carotene and the risk in adults of atherosclerosis, coronary heart disease, and certain cancers. FR 63, 6/22/98, 34092-34097	1. "Beta-carotene and other pro-vitamin A carotenoids can be converted to vitamin A in the body. Interest in the carotenoids has increased in recent years because of the accumulation of a large body of evidence that foods high in carotenoids are protective against a variety of epithelial cancers."	1. Contained in the "Nutrition Monitoring in the United States—An Update Report on Nutrition Monitoring" prepared for USDA and the Public Health Service of DHHS by LSRO of the FASEB.	The statement is not an "authoritative statement" because it indicates that the scientific evidence is preliminary or inconclusive, that it does not reflect the official policy of an appropriate scientific body, and that no appropriate scientific body has conducted a deliberative review of the scientific evidence.
	2. "The antioxidant nutrients found in plant foods (e.g., vitamin C, carotenoids, vitamin E, and certain minerals) are presently of great interest to scientists and the public because of their potentially beneficial role in reducing the risk of cancer and certain other chronic diseases."	2. From an electronic version of "Nutrition and Your Health: Dietary Guidelines for Americans," issued by DHHS and USDA and provided on the Internet.	FDA stated that the wording and context of the statement show that it is not an authoritative statement under section 403(r)(3)(C) of the Act.
	3. "If the findings hold up in further research, eating more vegetables rich in beta-carotene and related carotenoids– lutein and lycopene–may help people ward off a cold or flu as well as protect from cancer .... The findings also suggest that carotenoid-rich vegetables also stimulate the immune system."	3. Found in Human Nutrition (quarterly reports of selected research projects) issued by USDA's ARS and provided on the Internet.	USDA stated that the statement is not an "authoritative statement" of USDA because it was not based upon a deliberative review of the scientific evidence regarding a relationship between the nutrient and the disease in question.
	4. "This research involving cells provides data which supports the general hypothesis that beta-carotene and lutein	4. The statement is found in an interpretative summary of a research report	FDA concluded that the statement does not address a disease or health-related condition and therefore is

Adopted from the Federal Register (63 FR 34084-34115; (June 22, 1998)

Chart VI.F.3(c)
(Continued)

Claim	Statement	Authoritative Body	Basis for Denial
Health Claim: Antioxidant vitamin A and beta-carotene and the risk in adults of atherosclerosis, coronary heart disease, and certain cancers. FR 63, 6/22/98, 34092-34097 (Continued)	protect cells by serving as antioxidants."	from Technology Transfer Information Center, TEKTRAN of USDA/ARS entitled: "Beta-carotene and Lutein Protect the Plasma Membrane of HEPG2 Human Liver Cells Against Oxidant-induced Damage," and provided on the Internet.	not an "authoritative statement" under section 403(r)(3)(C) of the Act.
	5. "[Antioxidants] may help prevent disease. Antioxidants fight harmful molecules called oxygen free radicals, which are created by the body as cells go about their normal business of producing energy... [S]ome studies show that antioxidants may help prevent heart disease, some cancer, cataracts, and other health problems that are more common as people get older."	5. Found in "Life Extension: Science or Fiction?" that was provided on the Internet by the Administration on Aging and which includes statements from the "Age Page" of the National Institute on Aging" (an Institute of the NIH).	NIH stated that the statement is not an "authoritative statement" of NIH because it was prepared by an individual from the National Institute on Aging and is not based on a deliberative review of the scientific evidence regarding the nutrient-disease relationship in question.
	6. "As potent antioxidants, [lutein and lycopene] are thought to contribute to the lower rates of heart disease, cancer and other diseases of aging among populations that eat a lot of fruits and vegetables."	6. Found in "BHNRC Success Stories," provided on the Internet by USDA/ARS.	USDA stated that the statement is not an "authoritative statement" of USDA because it was not based upon a deliberative review of the scientific evidence regarding the nutrient-disease relationship in question.
	7. "Researchers also found more evidence suggesting that carotenes act as antioxidants to protect the body from harmful oxidation. Antioxidants	7. Found in Human Nutrition (quarterly reports of selected research projects), which is issued by USDA's ARS and provided on the	USDA stated that the statement is not an "authoritative statement" of USDA because it was not based upon a deliberative review of the scientific

Chart VI.F.3(c)
(Continued)

Chapter VI

Claim	Statement	Authoritative Body	Basis for Denial
Health Claim: Antioxidant vitamin A and beta-carotene and the risk in adults of atherosclerosis, coronary heart disease, and certain cancers. FR 63, 6/22/98, 34092-34097 (Continued)	are thought to help prevent heart attack, stroke and cancer. During the low-carotene stints, researchers recorded several biochemical signs of oxidative damage."	Internet.	evidence regarding the nutrient-disease relationship in question.
	8. "[H]igh dietary carotene and possibly vitamins C and E and folate are associated with reduced risk for cervical cancer."	8. Found in information on the Internet by the NCI, an Institute of NIH, in an article entitled: "Prevention of Cervical Cancer."	NIH stated that the statement was not an authoritative statement of NIH and does not reflect consensus within NIH.
	9. "[B]eta carotene or vitamin A supplements have reversed pre-cancerous conditions in people's mouths."	9. Found in Human Nutrition (quarterly reports of selected research projects), which is issued by the USDA's ARS and provided on the Internet.	USDA stated that the statement is not an authoritative statement of USDA because it was not based upon a deliberative review of the scientific evidence regarding the nutrient-disease relationship in question.
	10. "Carotenoids or other plants' components appear to boost the immune system."	10. Found in an interpretative summary of a research report from Technology Transfer Information Center, TEKTRAN of USDA/ARS and provided on the Internet.	FDA found that the statement is not an "authoritative statement" because it does not address a disease or health-related condition.
	11. "A wealth of epidemiological evidence has linked a high intake of green leafy and deep yellow vegetables—both rich in beta-carotene—with lower rates of many types of cancer.... Men over 65 who took a 50-	11. Found in Human Nutrition (quarterly reports of selected research projects), which is issued by the USDA's ARS and provided on the Internet.	USDA stated that the statement is not an "authoritative statement" of USDA because it was not based upon a deliberative review of the scientific evidence regarding the nutrient-disease relationship in question.

Adopted from the Federal Register (63 FR 34084-34115; (June 22, 1998)

Chart VI.F.3(c)
(Continued)

Claim	Statement	Authoritative Body	Basis for Denial
Health Claim: Antioxidant vitamin A and beta-carotene and the risk in adults of atherosclerosis, coronary heart disease, and certain cancers. FR 63, 6/22/98, 34092-34097 (Continued)	milligram beta-carotene supplement every other day during the 12-year study had natural killer cells that were more active than their counterparts who received a placebo. Natural killer cells—or NK cells—are the immune system's sentinels, ever on watch for viruses and cancer cells."		

Abbreviations used in chart:

FDA–Food and Drug Administration; CDC–Centers for Disease Control and Prevention; NIH–National Institutes of Health; NCI—National Cancer Institute; DHHS–Department of Health and Human Services; USDA–U.S. Department of Agriculture; ARS–Agricultural Research Service; BHNRC—Beltsville Human Nutrition Research Center; LSRO–Life Sciences Research Office; FASEB–Federation of American Societies for Experimental Biology; FR—*Federal Register*; FDAMA–Food and Drug Administration Modernization Act; and "the Act" —Federal Food, Drug, and Cosmetic Act.

Under the provisions of the Food and Drug Administration Modernization Act of 1997 (FDAMA), a manufacturer may submit to FDA a notification of a health claim based on an authoritative statement from an appropriate federal agency or the National Academy of Sciences (NAS). If FDA does not act to prohibit or modify such a claim within 120 days of receipt of the notification, the claim may be used.

HEALTH CLAIMS AUTHORIZED BASED ON AUTHORITATIVE STATEMENTS BY FEDERAL SCIENTIFIC BODIES			
APPROVED CLAIMS	**FOOD REQUIREMENTS**	**CLAIM REQUIREMENTS**	**MODEL CLAIM STATEMENTS**

Whole Grain Foods and Risk of Heart Disease and Certain Cancers Docket No. 99P-2209
 	
Updated 12/9/03 Docket No. 03Q-0547 (see listing VI.F.3(f))
	- Contains 51 percent or more whole grain ingredients by weight per RA, and
	
	- Dietary fiber content at least:
	
	  • 3.0 g per RA of 55 g
	
	  • 2.8 g per RA of 50 g
	
	  • 2.5 g per RA of 45 g
	
	  • 1.7 g per RA of 35 g
	
	- Low fat
		Required wording of the claim:
		
		" Diets rich in whole grain foods and other plant foods and low in total fat, saturated fat, and cholesterol may reduce the risk of heart disease and some cancers."
			NA

Potassium and the Risk of High Blood Pressure and Stroke Docket No. 00Q-1582	- Good source of potassium - Low sodium - Low total fat - Low saturated fat - Low cholesterol	*Required wording for the claim:* "Diets containing foods that are a good source of potassium and that are low in sodium may reduce the risk of high blood pressure and stroke."	NA

Adopted from FDA, CFSAN, Food Labeling Guide, September 1994 (Revised June 1999 and November 2002), and FDA, CFSAN, Health Claim Notification for Whole Grain Foods with Moderate Fat Content, December 9, 2003

Chapter VI

Chart VI.F.3(e)
General Criteria and Disqualifying Nutrient Levels

General Criteria All Claims Must Meet

- All information in one place without intervening material (Reference statement permitted).

- Only information on the value that intake or reduced intake, as part of a total dietary pattern, may have on a disease or health-related condition.

- Enables public to understand information provided and significance of information in the context of a total daily diet.

- Complete, truthful, and not misleading.

- Food contains, without fortification, 10% or more of the Daily Value for one of six nutrients (dietary supplements excepted):

Vitamin A	500 IU	Calcium	100 mg
Vitamin C	6 mg	Protein	5 g
Iron	1.8 mg	Fiber	2.5 g

- Not represented for infants or toddlers less than 2 years of age.

- Does not quantify any degree of risk reduction. Uses "may" or "might" to express relationship between substance and disease.

- Indicates disease depends on many factors.

- Food contains less than the specified levels of four disqualifying nutrients:

Disqualifying Nutrients	Foods	Main Dishes	Meal Products
Fat	13 g	19.5 g	26 g
Saturated Fat	4 g	6 g	8 g
Cholesterol	60 mg	90 mg	120 mg
Sodium	480 mg	720 mg	960 mg

Abbreviations: RA = reference amount, IU = International Units

Adopted from U.S. FDA, CFSAN, Food Labeling Guide, September 1994 (Editorial Revision June 1999 and November 2002)

| Chart VI.F.3(f) | **Chapter VI** |
| Listing of Specific FDAMA Claims | |

Specific FDAMA Claims

- **Choline**
  - <u>Nutrient Content Claims Notification for Choline Containing Foods.</u> August 30, 2001

- **Potassium and the Risk of High Blood Pressure and Stroke**
  - <u>Health Claim Notification for Potassium Containing Foods.</u> October 31, 2000

- **Whole Grain Foods and the Risk of Heart Disease and Certain Cancers**
  - <u>Health Claim Notification for Whole Grain Foods with Moderate Fat Content.</u> December 9, 2003
  - <u>Health Claim Notification for Whole Grain Foods.</u> July 8, 1999

- **Fluoridated Water**
  - <u>Health Claim Notification for Fluoridated Water and Reduced Risk of Dental Caries.</u> October 14, 2006

- **Saturated Fat, Cholesterol, and *Trans* Fat, and the Risk of Heart Disease**
  - <u>Health Claim Notification for Saturated Fat, Cholestrol, and *Trans* Fat and Reduced Risk of Heart Disease.</u> November 15, 2006

(a) Summary Comments

FDA's 2003 *Consumer Health Information for Better Nutrition Initiative* provides for the use of qualified health claims when there is emerging evidence for a relationship between a food, food component, or dietary supplement and reduced risk of a disease or health-related condition. In this case, the evidence is not well enough established to meet the significant scientific agreement standard required for FDA to issue an authorizing regulation. Qualifying language is included as part of the claim to indicate that the evidence supporting the claim is limited. Both conventional foods and dietary supplements may use qualified health claims. FDA uses its enforcement discretion for qualified health claims after evaluating and ranking the quality and strength of the totality of the scientific evidence. Although FDA's "enforcement discretion" letters are issued to the petitioner requesting the qualified health claim, the qualified claims are available for use on any food or dietary supplement product meeting the enforcement discretion conditions specified in the letter. FDA has prepared a guide on interim procedures for qualified health claims (see VI.F.4(b)) and, an internal evidence-based ranking system for scientific data for ranking the strength of the evidence supporting a qualified claim, see VI.F.4(c). Example of FDA's decision letters is found at VI.F.4(d).

Guidance VI.F.4(b)
Interim Procedures for Qualified Health Claims in the Labeling of Conventional Human Food and Dietary Supplements

Chapter VI

Guidance for Industry and FDA

Interim Procedures for Qualified Health Claims in the Labeling of Conventional Food and Human Dietary Supplements

Conventional Human Food and Dietary Supplements

This guidance is intended to notify the public of interim procedures that the Food and Drug Administration (FDA) is implementing for petitioners who submit qualified health claim petitions to the agency. This guidance describes the procedures that FDA intends to use, on an interim basis, to respond to qualified health claim petitions until the agency can promulgate regulations under notice-and-comment rulemaking; it also provides a linkage between the ranking of scientific evidence and the wording of qualified health claims. In addition, this guidance updates the agency's approach outlined in December 2002 (Guidance for Industry: Qualified Health Claims in the Labeling of Conventional Foods and Dietary Supplements) and the agency's approach to implementing *Pearson v. Shalala* (164 F.3d 650 (D.C. Cir. 1999)) to include conventional foods. This guidance does not apply to unqualified health claims, which must meet the "Significant Scientific Agreement" (SSA) standard.

FDA's guidance documents, including this guidance, do not establish legally enforceable responsibilities. Instead, guidances describe the Agency's current thinking on a topic and should be viewed only as recommendations, unless specific regulatory or statutory requirements are cited. The use of the word *should* in Agency guidances means that something is suggested or recommended, but not required.

FDA intends to use the following interim procedures to ensure that its premarket review is consistent with the spirit of the Nutrition Labeling and Education Act and the First Amendment. FDA will continue to evaluate unqualified health claims under its current regulatory process and standard for significant scientific agreement *(21 CFR 101.14 and 101.70)*.

Criteria for Exercise of Enforcement Discretion

FDA plans to establish criteria for considering exercising enforcement discretion for qualified health claims based on the extent to which the totality of the publicly available evidence supports the claim (see Guidance for Industry and FDA: Interim Evidence-based Ranking System for Scientific Data). Different levels of evidence will result in different qualifying language as described in Table 1, which provides standardized language for the B, C, and D categories to be used as part of the qualifying language for qualified health claims until consumer research is complete.

Table 1. Standardized Qualifying Language for Qualified Health Claims.		
Scientific Ranking*	**FDA Category**	**Appropriate Qualifying Language****
Second Level	B	... "although there is scientific evidence supporting the claim, the evidence is not conclusive."
Third Level	C	"Some scientific evidence suggests ... however, FDA has determined that this evidence is limited and not conclusive."
Fourth Level	D	"Very limited and preliminary scientific research suggests... FDA concludes that there is little scientific evidence supporting this claim."

*From Guidance for Industry and FDA: Interim Evidence-based Ranking System for Scientific Data.

**The language reflects wording used in qualified health claims as to which the agency has previously exercised enforcement discretion for certain dietary supplements. During this interim period, the precise language as to which the agency considers exercising enforcement discretion may vary depending on the specific circumstances of each case.

Comments and suggestions regarding this document may be submitted at any time. Submit comments to the Division of Dockets Management (HFA-305), Food and Drug Administration, 5630 Fishers Lane, rm. 1061, Rockville, MD 20852. All comments should be identified with the docket number listed in the notice of availability that publishes in the *Federal Register*.

For questions regarding this draft document contact the Center for Food Safety and Applied Nutrition (CFSAN) at 301-436-1450.

Copies of the entire document are available from Office of Nutritional Products, Labeling, and Dietary Supplement, Division of Nutrition Programs and Labeling HFS-800, Center for Food Safety and Applied Nutrition, Food and Drug Administration, 5100 Paint Branch Parkway, College Park, MD 20740 (Tel) 301-436-1450, http://www.cfsan.fda.gov/guidance.html

Adopted from DHHS, FDA, CFSAN, Guidance for Industry and FDA, Interim Procedures for Qualified Health Claims in the Labeling of Conventional Human Food and Human Dietary Supplements, July 2003

177

Guidance for Industry and FDA
Interim Evidence-based Ranking System for Scientific Data
GUIDANCE

Contains Nonbinding Recommendations

Guidance for Industry and FDA
Interim Evidence-based Ranking System for Scientific Data

This guidance is intended to notify the public of the Food and Drug Administration's (FDA) interim evidence-based ranking system that is a process designed to lay a foundation for a more detailed system to be used permanently. This guidance describes a process that FDA intends to use, on an interim basis, to evaluate and rank the scientific evidence in support of a substance/disease relationship that is the subject of a qualified health claim until the agency can promulgate regulations under notice-and-comment rulemaking. Based on this process, the agency will categorize the qualified health claim into one of three levels (i.e., a "B", "C", or "D" level). This guidance does not apply to unqualified health claims, which must meet the "Significant Scientific Agreement" (SSA) standard.

FDA's guidance documents, including this guidance, do not establish legally enforceable responsibilities. Instead, guidances describe the Agency's current thinking on a topic and should be viewed only as recommendations, unless specific regulatory or statutory requirements are cited. The use of the word *should* in Agency guidances means that something is suggested or recommended, but not required.

The result of the evidence-based rating system will be a statement describing the nature of the evidence and the rationale for linking a substance to a disease/health-related condition with a ranking as to the strength of the scientific evidence in support of that relationship. The process for arriving at the rank of the evidence to support the substance/disease relationship is illustrated in **Table 1**. The rank will be supported by:

Table 1. Overview of the evidence-based rating system for evaluating the substance/disease relationship that is the subject of a qualified health claim.
There are six steps to evaluating the strength of the scientific evidence in support of a qualified health claim.
Step One. A proposed relationship between a substance and a disease or health-related condition is identified.
Step Two. Individual studies are identified that are pertinent to the substance/disease relationship.
Step Three. Individual studies are classified according to study design type. Different design types are graded higher than others, based on their ability to minimize bias. Thus assignment of a study design automatically provides a rating.
Step Four. Individual studies are assigned a designator of +, Ø, -, or N/A to reflect the study quality. (The general criteria for quality determination are described in this guidance).
Step Five. The strength of the scientific evidence in support of the substance/disease relationship is given a rank. This rank is determined taking into account the quantity, consistency, and relevance to disease risk reduction of the *aggregate* of the studies.
Step Six. The rank is reported.

Chapter VI

Copies of the entire document are available from, the Office of Nutritional Products, Labeling, and Dietary Supplements, Division of Nutrition Programs and Labeling HFS-800, Center for Food Safety and Applied Nutrition, Food and Drug Administration, 5100 Paint Branch Parkway, College Park, MD 20740, (Tel) 301-436-1450, http://www.cfsan.fda.gov/guidance.html

Adopted from DHHS, FDA, CFSAN, Guidance for Industry and FDA, Interim Procedures
for Qualified Health Claims in the Labeling of Conventional Human Food and Human
Dietary Supplements, July 2003

179

Decision Letter Regarding Dietary Supplement Health Claim for Antioxidant Vitamins and Risk of Certain Cancers

Jonathan W. Emord
Emord and Associates, P.C.
Suite 600
1050 17th Street, N.W.
Washington, D.C. 20036

Dear Mr. Emord:

This letter is a follow-up to the Food and Drug Administration's February 11, 2003, letter, pursuant to the opinion and order issued December 26, 2002, by the U.S. District for the District of Columbia in *Whitaker, et al. v. Thompson, et al.*, Civil No. 01-1539, and your response dated February 13, 2003. As you know, the United States withdrew its notice of appeal in this matter on March 28. The purpose of this letter is to formalize the contingent understanding we reached in mid-February.

In its opinion and order, the Court instructed FDA to draft one or more "short, succinct, and accurate disclaimers" for the health claim: "Consumption of antioxidant vitamins may reduce the risk of certain kinds of cancer." Slip Op. at 37; see Order at 1-2.

As we explained in our February 11 letter, FDA considered the two disclaimers suggested by the Court, as well as a number of others, and concluded the following three alternative disclaimers best meet the criteria specified in the Court's decision:

1. Some scientific evidence suggests that consumption of antioxidant vitamins may reduce the risk of certain forms of cancer. However, FDA has determined that this evidence is limited and not conclusive.

2. Some scientific evidence suggests that consumption of antioxidant vitamins may reduce the risk of certain forms of cancer. However, FDA does not endorse this claim because this evidence is limited and not conclusive.

3. FDA has determined that although some scientific evidence suggests that consumption of antioxidant vitamins may reduce the risk of certain forms of cancer, this evidence is limited and not conclusive.

Your February 13 letter indicated that your clients accept these disclaimers, and that the various petitioners in this matter wish to have the option of using any of three disclaimers on their products with the antioxidant vitamin claim.

FDA intends to exercise its enforcement discretion with respect to antioxidant vitamin dietary supplements containing vitamin E and/or vitamin C when: (1) one of the above disclaimers is placed immediately adjacent to and directly beneath the antioxidant vitamin claim, with no intervening material, in the same size, typeface, and contrast as the claim itself; and (2) the supplement does not recommend or suggest in its labeling, or under ordinary conditions of use, a daily intake exceeding the Tolerable Upper Intake Level established by the Institute of Medicine (IOM) of 2,000 mg per day for vitamin C and 1,000 mg per day for vitamin E (see May 4, 2001, letter at 4-6 and references cited therein).

Antioxidant vitamin supplements bearing the claim and one of the disclaimers are still required to meet all applicable statutory and regulatory requirements under the Federal Food, Drug, and Cosmetic Act, including the applicable requirements for health claims.

Sincerely,

Christine L. Taylor, Ph.D.
Director Office of Nutritional Products, Labeling and Dietary Supplements
Center for Food Safety and Applied Nutrition

Chart VI.F.4(e)
Examples of Permitted
Qualified Health Claims

Qualified Claims About Cancer Risk			
Item	**Eligible Foods**	**Required Claim**	**Conditions**
Selenium & Cancer Docket No. 02P-0457	Dietary supplements containing selenium	(1) Selenium may reduce the risk of certain cancers. Some scientific evidence suggests that consumption of selenium may reduce the risk of certain forms of cancer. However, FDA has determined that this evidence is limited and not conclusive. *or,* (2) Selenium may produce anticarcinogenic effects in the body. Some scientific evidence suggests that consumption of selenium may produce anticarcinogenic effects in the body. However, FDA has determined that this evidence is limited and not conclusive.	The disclaimer (i.e., Some scientific evidence suggests...) is placed immediately adjacent to and directly beneath the claim (i.e., Selenium may reduce the risk), with no intervening material, in the same size, typeface, and contrast as the claim itself. The supplement does not recommend or suggest in its labeling, or under ordinary conditions of use, a daily intake exceeding the Tolerable Upper Intake Level established by the National Academy of Sciences/Institute of Medicine for selenium (400 micrograms per day). The claim meets all general health claim requirements of *21 CFR 101.14, except* for the requirement that the evidence for the claim meets the significant scientific agreement standard and be made in accordance with an authorizing regulation. Paragraph *101.14(d)(2)(vii)* requires that the dietary supplement bearing the claim meet the nutrient content claim definition for high (i.e., 20% or more of the Daily Value (DV) per RACC). 20% DV for selenium is 14 micrograms.

Chart VI.F.4(e)	Chapter VI
(Continued)	

Item	Eligible Foods	Required Claim	Conditions
Antioxidant Vitamins & Cancer Docket No. 91N-0101	Dietary supplements containing vitamin E and/or vitamin C	(1) Some scientific evidence suggests that consumption of antioxidant vitamins may reduce the risk of certain forms of cancer. However, FDA has determined that this evidence is limited and not conclusive. *or,* (2) Some scientific evidence suggests that consumption of antioxidant vitamins may reduce the risk of certain forms of cancer. However, FDA does not endorse this claim because this evidence is limited and not conclusive. *or,* (3) FDA has determined that although some scientific evidence suggests that consumption of antioxidant vitamins may reduce the risk of certain forms of cancer, this evidence is limited and not conclusive.	The disclaimer (i.e., ...evidence is limited and not conclusive) is placed immediately adjacent to and below the claim, with no intervening material, in the same size, typeface, and contrast as the claim itself. The supplement does not recommend or suggest in its labeling, or under ordinary conditions of use, a daily intake exceeding the Tolerable Upper Intake Levels established by the Institute of Medicine for vitamin C (2,000 mg per day) or for vitamin E (1,000 mg per day). The claim meets all *21 CFR 101.14* general health claim requirements, *except* for the requirements that the claim meet the significant scientific agreement standard and be made in accordance with an authorizing regulation. Paragraph 101.14(d)(2)(vii) requires that the food bearing the claim meet the nutrient content claim definition for *high* (i.e., 20% or more of the Daily Value (DV) per RACC). 20% DV for vitamin C is 12 mg; 20% DV for vitamin E is 6 IU.

Qualified Claims About Cardiovascular Disease Risk

Item	Eligible Foods	Required Claim	Conditions
Nuts & Heart Disease Docket No. 02P-0505	(1) *Whole or chopped nuts* listed below that are raw, blanched, roasted, salted, and/or lightly coated and/or flavored; any fat or	Scientific evidence suggests but does not prove that eating 1.5 ounces per day of most nuts [such as *name of*	*Whole or chopped nuts* The claim meets all *21 CFR 101.14* general health claim require- ments, except for:

Chart VI.F.4(e)
(Continued)

Item	Eligible Foods	Required Claim	Conditions
Nuts & Heart Disease Docket No. 02P-0505 (Continued)	carbohydrate added in the coating or flavoring must meet the § 101.9(f)(1) definition of an insignificant amount. 2) *Nut-containing products* other than whole or chopped nuts that contain at least 11 g of one or more of the nuts listed below per RACC. (3) Types of nuts eligible for this claim are restricted to almonds, hazelnuts, peanuts, pecans, some pine nuts, pistachio nuts, and walnuts. Types of nuts on which the health claim may be based is restricted to those nuts that were specifically included in the health claim petition, but that do not exceed 4 g saturated fat per 50 g of nuts	*specific nut*] as part of a diet low in saturated fat and cholesterol may reduce the risk of heart disease. [See nutrition information for fat content.] *Notes: The bracketed phrase naming a specific nut is optional. The bracketed fat content disclosure statement is applicable to a claim made for whole or chopped nuts, but not a claim made for nut-containing products.*	(1) the requirement that the claim meet the significant scientific agreement standard and be made in accordance with an authorizing regulation; (2) the § 101.14(a)(4) requirement that the food comply with the total fat disqualifying level; and (3) for walnuts only, the § 101.14(e)(6) requirement that the food contain a minimum of 10 percent of the Daily Value per RACC of vitamin A, vitamin C, iron, calcium, protein, or dietary fiber. Where the claim is used on whole or chopped nuts, the disclosure statement (see nutrition information...) must be placed immediately adjacent to and directly beneath the claim, with no intervening material, in the same size, typeface, and contrast as the claim itself. Nuts bearing the claim must comply with the § 101.14(a)(4) saturated fat disqualifying level (4 g saturated fat per 50 g nuts). *Nut-containing products* The claim meets all *21 CFR 101.14* general health claim requirements, *except* for the requirement that the claim meet the significant scientific agreement standard and be made in accordance with an authorizing regulation

Chart VI.F.4(e)
(Continued)

Item	Eligible Foods	Required Claim	Conditions
Nuts & Heart Disease Docket No. 02P-0505 (Continued)			Nut-containing products bearing the claim must comply with all the § 101.14(a)(4) disqualifying levels which are 13 g total fat, 4 g saturated fat, 60 mg of cholesterol, and 480 mg of sodium per RACC. The claim applies only to types of nuts that do not exceed the § 101.14(a)(4) disqualifying nutrient level for saturated fat (4 g saturated fat per 50 g nuts). Nut-containing products bearing the claim must comply with the § 101.62(c)(2) definition of a *low saturated fat food* and the § 101.62(d)(2) definition of a *low cholesterol food*. Nut-containing products bearing the claim must comply with the § 101.14(e)(6) requirement that the food contain a minimum of 10 percent of the Daily Value per RACC of vitamin A, vitamin C, iron, calcium, protein, or dietary fiber prior to any nutrient addition.
Walnuts & Heart Disease Docket No. 02P-0292	Whole or chopped walnuts	(1) Supportive but not conclusive research shows that eating 1.5 ounces per day of walnuts as part of a diet low in saturated fat and cholesterol may reduce the risk of heart disease. See nutrition information for fat content. *or,* (2) Scientific evidence suggests but does not	The claim meets all *21 CFR 101.14* general health claim requirements, *except* for: (1) the requirement that the claim meet the significant scientific agreement standard and be made in accordance with an authorizing regulation; authorizing regulation; (2) the § 101.14(a)(4)

Chart VI.F.4(e)
(Continued)

Item	Eligible Foods	Required Claim	Conditions
Walnuts & Heart Disease Docket No. 02P-0292 (Continued)		prove that eating 1.5 ounces per day of most nuts, such as walnuts, as part of a diet low in saturated fat and cholesterol may reduce the risk of heart disease. See nutrition information for fat content.	requirement that the food comply with the total fat disqualifying level; and (3) *the § 101.14(e)(6)* requirement that the food contain a minimum of 10 percent of the Daily Value per RACC of vitamin A, vitamin C, iron, calcium, protein, or dietary fiber.
			The disclosure statement (i.e., See nutrition information...) must be placed immediately adjacent to and directly beneath the claim, with no intervening material, in the same size, typeface, and contrast as the claim itself.
Omega-3 Fatty Acids & Coronary Heart Disease Docket No. 91N-0103	Dietary supplements containing the omega-3 long chain polyunsaturated fatty acids eicosapentaenoic acid (EPA) and/or docosahexaenoic acid (DHA)	Consumption of omega-3 fatty acids may reduce the risk of coronary heart disease. FDA evaluated the data and determined that, although there is scientific evidence supporting the claim, the evidence is not conclusive.	The claim does not recommend or suggest in its labeling, or under ordinary conditions of use, a daily intake exceeding 2 grams per day of EPA and DHA. FDA encourages manufacturers to limit the products that bear the qualified claim to a daily intake of 1 gram of omega-3 fatty acids or below.
			Dietary supplements exceeding the § 101.14(a)(4) total fat disqualifying level must have a disclosure statement (i.e., "see nutrition information for fat content") immediately adjacent to the claim.
			The claim meets all *21 CFR 101.14* general health claim

Chart VI.F.4(e)
(Continued)

Chapter VI

Item	Eligible Foods	Required Claim	Conditions
Omega-3 Fatty Acids & Coronary Heart Disease Docket No. 91N-0103 (Continued)			requirements, *except* for: (1) the requirements that the claim meet the significant scientific agreement standard and be made in accordance with an authorizing regulation, and (2) the claim specify the daily dietary intake necessary to achieve the claimed effect. The claim may not suggest a level of omega-3 fatty acids as being useful in achieving the claimed effect.
B Vitamins & Vascular Disease Docket No. 99P-3029	Dietary supplements containing vitamin B6, B12, and/or folic acid	As part of a well-balanced diet that is low in saturated fat and cholesterol, Folic Acid, Vitamin B6 and Vitamin B12 may reduce the risk of vascular disease. FDA evaluated the above claim and found that, while it is known that diets low in saturated fat and cholesterol reduce the risk of heart disease and other vascular diseases, the evidence in support of the above claim is inconclusive.	The disclaimer (i.e., FDA evaluated the above claim...) must be immediately adjacent to and directly beneath the first claim (i.e., As part of a well-balanced diet...) with no intervening material that separates the claim from the disclaimer, and the second sentence must be in the same size, type face and contrast as the first sentence. Products that contain more than 100 percent of the Daily Value (DV) of folic acid (400 micrograms), when labeled for use by adults and children 4 or more years of age, must identify the safe upper limit of daily intake with respect to the DV. The folic acid safe upper limit of daily intake value of 1,000 micrograms (1 mg) may be included in parentheses. The claim meets all *21 CFR 101.14* general health claim requirements, *except* for: (1) the requirement

Chart VI.F.4(e)
(Continued)

Item	Eligible Foods	Required Claim	that the claim meet the Conditions
B Vitamins & Vascular Disease Docket No. 99P-3029 (Continued)			significant scientific agreement standard and be made in accordance with an authorizing regulation, and (2) the requirement that the claim specify the daily dietary intake necessary to achieve the claimed effect. The claim may not suggest a level of vitamins B6, B12, and/or folic acid as being useful in achieving the claimed effect. Dietary supplements containing folic acid must meet the United States Pharmacopeia (USP) standards for disintegration and dissolution, except that if there are no applicable USP standards, the folate in the dietary supplement shall be shown to be bioavailable under the conditions of use stated on the product label.

Qualified Claims About Cognitive Function

Item	Eligible Foods	Required Claim	Conditions
Phosphatidylserine & Cognitive Dysfunction and Dementia Docket No. 02P-0413	Dietary supplements containing soy-derived phosphatidylserine	(1) Consumption of phosphatidylserine may reduce the risk of dementia in the elderly. Very limited and preliminary scientific research suggests that phosphatidylserine may reduce the risk of dementia in the elderly. FDA concludes that there is little scientific evidence supporting this claim. *or,* (2) Consumption of phosphatidylserine may reduce the risk of	The disclaimer (i.e., Very limited and preliminary scientific research...) is placed immediately adjacent to and directly beneath the claim (i.e., Phosphatidylserine may reduce...), with no intervening material, in the same size, typeface, and contrast as the claim itself. The claim meets all *21 CFR 101.14* general health claim requirements, *except* for:

Chart VI.F.4(e)
(Continued)

Chapter VI

Item	Eligible Foods	Required Claim	Conditions
Phosphatidylserine & Cognitive Dysfunction and Dementia Docket No. 02P-0413 (Continued)		cognitive dysfunction in the elderly. Very limited and preliminary scientific research suggests that phosphatidylserine may reduce the risk of cognitive dysfunction in the elderly. FDA concludes that there is little scientific evidence supporting this claim.	(1) the requirement that the claim meet the significant scientific agreement standard and be made in accordance with an authorizing regulation, and (2) the claim specify the daily dietary intake necessary to achieve the claimed effect. The claim may not suggest a level of phosphatidylserine as being useful in achieving the claimed effect. The soy-derived phosphatidylserine used is of very high purity.

Qualified Claims About Neural Tube Birth Defects

Item	Eligible Foods	Required Claim	Conditions
0.8 mg Folic Acid & Neural Tube Birth Defects Docket No. 91N-100H	Dietary supplements containing folic acid	0.8 mg folic acid in a dietary supplement is more effective in reducing the risk of neural tube defects than a lower amount in foods in common form. FDA does not endorse this claim. Public health authorities recommend that women consume 0.4 mg folic acid daily from fortified foods or dietary supplements or both to reduce the risk of neural tube defects.	The disclaimer (i.e., FDA does not endorse this claim...) is placed immediately adjacent to and directly beneath the claim (i.e., 0.8 mg folic acid ...), with no intervening material, in the same size, typeface, and contrast as the claim. The claim meets all *21 CFR 101.14* general health claim requirements, *except* for the requirements that the claim meet the significant scientific agreement standard and be made in accordance with an authorizing regulation. **Note:** there also is a folic acid/neural tube defect health claim authorized by regulation (see *21 CFR 101.79*).

Abbreviations: RACC - reference amount customarily consumed per eating occasion, as defined in 21 CFR 101.12

Chapter VI

5. Statement Made Concerning the Effect of a Product on the Structure or Function of the Body

(a) Summary Comments

Structure/function claims have historically appeared on labels of conventional foods and dietary supplements as well as drugs. When a conventional food's label or labeling contained statements about the effect of a substance on the structure or function of the body such statements could not claim to diagnose, mitigate, treat, cure, or prevent a disease; it could not be false or misleading; the claimed effect must have been achieved through nutritive value; and there are no specific regulations for such claims. As a legal matter, a structure/function claim on a conventional food that is not achieved through nutritive value, or if it claimed to treat or mitigate a disease, such a product would be subject to regulation as a drug under section 201 (g)(1) of the Act. However, the Dietary Supplement Health and Education Act of 1994 (DSHEA) established some special regulatory procedures for such claims for dietary supplements. In implementing these requirements of DSHEA on January 6, 2000 (65 FR 999), the FDA published a regulation revising section (f) and (g) of *21 CFR 101.93* (see regulation VI.F.5(c)), and provided examples of how the agency would distinguish between allowable structure/function claims and disease claims (see chart VI.5(e)). In the preamble to that final regulation, FDA stated that the agency is likely to interpret the dividing line between structure/function claims and disease claims in a similar manner for conventional foods as for dietary supplements. Therefore, the following information concerning evaluating claims for appropriateness as structure/function claims for dietary supplements is presented here to serve as a guide in the review of claims on food labels.

Adopted from the Federal Register, Vol. 65, pp. 999-1050, 1/6/00; and FDA, CFSAN, ONPLDS, Claims That Can Be Made for Conventional Foods and Dietary Supplements, March 20, 2001 and revised October 2001 and September 2003

Regulation VI.F.5(b)
Notification Procedures for Certain Types
of Statements on Dietary Supplements
(21 CFR 101.93(a) through (e))

Chapter VI

Notification procedures for certain types of statements on dietary supplements.
(21 CFR 101.93 (a)–(e))

(a) (1) No later than 30 days after the first marketing of a dietary supplement that bears one of the statements listed in section 403(r)(6) or the Federal Food, Drug, and Cosmetic Act, the manufacturer, packer, or distributor of the dietary supplement shall notify the Office of Nutritional Products, Labeling and Dietary Supplements (HFS-810), Center for Food Safety and Applied Nutrition, Food and Drug Administration, 5100 Paint Branch Pkwy., College Park, MD 20740, that it has included such a statement on the label or in the labeling of its product. An original and two copies of this notification shall be submitted.

(2) The notification shall include the following:

(i) The name and address of the manufacturer, packer, or distributor of the dietary supplement that bears the statement;

(ii) The text of the statement that is being made;

(iii) The name of the dietary ingredient or supplement that is the subject of the statement, if not provided in the text of the statement; and

(iv) The name of the dietary supplement (including brand name), if not provided in response to paragraph (a)(2)(iii) on whose label, or in whose labeling, the statement appears.

(3) The notice shall be signed by a responsible individual or the person who can certify the accuracy of the information presented and contained in the notice. The individual shall certify that the information contained in the notice is complete and accurate, and that the notifying firm has substantiation that the statement is truthful and not misleading.

(b) *Disclaimer*—The requirements in this section apply to the label or labeling of dietary supplements where the dietary supplement bears a statement that is provided for by section 403(r)(6) of the Federal Food, Drug, and Cosmetic Act (the act), and the manufacturer, packer, or distributor wishes to take advantage of the exemption to section 201(g)(1)(C) of the act that is provided by compliance with section 403(r)(6) of the act.

(c) *Text for disclaimer.*

(1) Where there is one statement, the disclaimer shall be placed in accordance with paragraph (d) of this section and shall state:
This statement has not been evaluated by the Food and Drug Administration. This product is not intended to diagnose, treat, cure, or prevent any disease.

(2) Where there is more than one such statement on the label or in the labeling, each statement shall bear the disclaimer in accordance with paragraph (c)(1) of this section, or a plural disclaimer may be placed in accordance with paragraph (d) of this section and shall state:
These statements have not been evaluated by the Food and Drug Administration. This product is not intended to diagnose, treat, cure, or prevent any disease.

(d) *Placement*—The disclaimer shall be placed adjacent to the statement with no intervening material or linked to the statement with a symbol (e.g., an asterisk) at the end of each such statement that refers to the same symbol placed adjacent to the disclaimer specified in paragraphs (c)(1) or (c)(2) of this section. On product labels and in labeling (e.g., pamphlets, catalogs), the disclaimer shall appear on each panel or page where there is such a statement. The disclaimer shall be set off in a box where it is not adjacent to the statement in question.

(e) *Typesize*—The disclaimer in paragraph (c) of this section shall appear in boldface type in letters of a typesize no smaller than one-sixteenth inch.

Adopted from 21 CFR Parts 100 through 169 (4/1/06 Edition)

Chapter VI

Regulation VI.F.5(c)
Certain Types of Statements on Dietary Supplements
(21 CFR 101.93(f) through (g))

Certain types of statements on dietary supplements. *(21 CFR 101.93(f) & (g))*

*****

(f) *Permitted structure/function statements.*

(1) Dietary supplement labels or labeling may, subject to the requirements of this section, bear statements that describe the role of a nutrient or dietary ingredient intended to affect the structure or function in humans or that characterize the documented mechanism by which a nutrient or dietary ingredient acts to maintain such structure or function, but may not bear statements that are disease claims under paragraph (g) of this section.

(g) *Disease claims.*

(1) *Definition of disease.* For purposes of 21 U.S.C. 343(r)(6), a "disease" is any deviation from, impairment of, or interruption of the normal structure or function of any part, organ, or system (or combination thereof) of the body that is manifested by a characteristic set of one or more signs or symptoms, including laboratory or clinical measurements that are characteristic of a disease.

(2) *Disease Claims.* FDA will find that a statement about a product claims to diagnose, mitigate, treat, cure, or prevent disease (other than a classical nutrient deficiency disease) under section 403(r)(6) of the act if it meets one or more of the criteria listed in this paragraph (g)(2). In determining whether a statement is a disease claim under these criteria, FDA will consider the context in which the claim is presented. A statement claims to diagnose, mitigate, treat, cure, or prevent disease if it claims, explicitly or implicitly, that the product:

(i) Has an effect on a specific disease or class of diseases;

(ii) Has an effect, using scientific or lay terminology, on one or more signs or symptoms that are recognizable to health care professionals or consumers as being characteristic of a specific disease or of a number of different specific diseases;

(iii) Has an effect on a consequence of a natural state that presents a characteristic set of signs or symptoms recognizable to health care professionals or consumers as constituting an abnormality of the body;

(iv) Has an effect on disease through one or more of the following factors:

A. The name of the product;

B. A statement about the formulation of the product, including a claim that the product contains an ingredient that has been regulated by FDA as a drug and is well known to consumers for its use in preventing or treating a disease;

C. Citation of the title of a publication or reference, if the title refers to a disease use;

D. Use of the term "disease" or "diseased"; or

E. Use of pictures, vignettes, symbols, or other means;

(v) Belongs to a class of products that is intended to diagnose, mitigate, treat, cure, or prevent a disease;

(vi) Is a substitute for a product that is a therapy for a disease;

(vii) Augments a particular therapy or drug action;

(viii) Has a role in the body's response to a disease or to a vector of disease;

(ix) Treats, prevents, or mitigates adverse events associated with a therapy for a disease and manifested by a characteristic set of signs or symptoms; or

(x) Otherwise suggests an effect on a disease or diseases.

Regulation VI.F.5(d)
**Other Evidence that the Intended Use of a Product
is for the Diagnosis, Cure, Mitigation, Treatment,
or Prevention of a Disease**
(21 CFR 201.128)

Chapter VI

Meaning of "intended uses."
(21 CFR 201.128)

The words *intended uses* or words of similar import in *§§ 201.5, 201.115, 201.117, 201.119, 201.120,* and *201.122* refer to the objective intent of the persons legally responsible for the labeling of drugs. The intent is determined by such persons' expressions or may be shown by the circumstances surrounding the distribution of the article. This objective intent may, for example, be shown by labeling claims, advertising matter, or oral or written statements by such persons or their representatives. It may be shown by the circumstances that the article is, with the knowledge of such persons or their representatives, offered and used for a purpose for which it is neither labeled nor advertised. The intended uses of an article may change after it has been introduced into interstate commerce by its manufacturer. If, for example, a packer, distributor, or seller intends an article for different uses than those intended by the person from whom he received the drug, such packer, distributor, or seller is required to supply adequate labeling in accordance with the new intended uses. But if a manufacturer knows, or has knowledge of facts that would give him notice, that a drug introduced into interstate commerce by him is to be used for conditions, purposes, or uses other than the ones for which he offers it, he is required to provide adequate labeling for such a drug which accords with such other uses to which the article is to be put.

Chart VI. F.5(e)
Examples of Disease Claims and Structure/Function Claims for Dietary Supplements
(21 CFR 101.93(g))

Disease Claims (Not Permitted on Dietary Supplement Labels)	Structure/Function Claims (Permitted on Dietary Supplement Labels)
§101.93(g)(2)(i): Statement that explicitly or implicitly claims an effect on a specific disease or class of diseases • (e.g., "Protective against the development of cancer," "reduces the pain and stiffness associated with arthritis," "decreases the effects of alcohol intoxication," "alleviates chronic constipation", "promotes low blood pressure"). • (e.g., Implied disease claims that convey prevention or treatment of a specific disease or class of diseases without actually mentioning the name of the disease which follows-"relieves crushing chest pain"-angina or heart attack, "prevents bone fragility in post-menopausal women"-osteoporosis, "improves joint mobility and reduces joint inflammation and pain"-rheumatoid arthritis, "heals stomach or duodenal lesions and bleeding"-ulcers, "anticonvulsant"-epilepsy, "relief of bronchospasm"-asthma, "preventing wasting in persons with weak immune system"-(AIDS) acquired immune deficiency syndrome, "prevents irregular heartbeat"-arrhythmias, "controls blood sugar in persons with insufficient insulin"-diabetes, "prevents the spread of neoplastic cells"-prevention of cancer metastases, "antibiotic"-infections, "herbal Prozac"-depression).	Statement that does not explicitly or implicitly claim an effect on a specific disease or class of diseases • (e.g., "Helps promote urinary tract health," "helps maintain cardiovascular function and a healthy circulatory system," "helps maintain intestinal flora," "promotes relaxation", "for relief of occasional constipation").
§101.93(g)(2(ii): Statement that explicitly or implicitly claims an effect (using scientific or lay terminology) on one or more signs or symptoms that are recognizable to health care professionals or consumers as being characteristic of a specific disease or of a number of diseases • (e.g. "Improves urine flow in men over 50 years old"-characteristic symptoms of benign prostatic hypertrophy) • (e.g., "Lowers cholesterol"-characteristic sign of hypercholesterolemia)	Statement of an effect on symptoms that are not recognizable as characteristic of a specific disease or of a number of diseases • (e.g., "Reduces stress and frustration," "improves absentmindedness") • (e.g., "Helps maintain cholesterol levels that are already within the normal range")

Chart VI. F.5(e)	Chapter VI
(Continued)	

Disease Claims (Not Permitted on Dietary Supplement Labels)	Structure/Function Claims (Permitted on Dietary Supplement Labels)
§101.93(g)(2(ii) (continued) • (e.g., "Reduces joint pain"-characteristic symptom of arthritis) • (e.g., "Relieves headache"-characteristic symptom of migraine or tension headache) • (e.g., "Maintains healthy lungs in smokers") Statements should not be made that products "restore" normal or "correct" abnormal function when the abnormality implies the presence of disease. An example might be a claim to "restore" normal blood pressure when the abnormality implies hypertension.	• (e.g., "Helps support cartilage and joint function") • (e.g., "For relief of occasional headache") • (e.g., "Maintains healthy lung functions") • (e.g., "Helps maintain regularity") Statement that a substance helps maintain normal function, if the context does not suggest treatment or prevention of disease
§101.93(g)(2)(iii): Statement that explicitly or implicitly claims an effect on an abnormal condition associated with a natural state or process, if the abnormal condition is uncommon or can cause significant or permanent harm • The following are examples of conditions that would be disease claims: Toxemia of pregnancy; hyperemesis gravidarum; acute psychosis of pregnancy; osteoporosis; Alzheimer's disease, and other senile dementias; glaucoma; arteriosclerotic diseases of coronary, cerebral or peripheral blood vessels; cystic acne; and severe depression associated with the menstrual cycle. "Helps to maintain normal urine flow in men over 50 years old" is an implied disease claim.	Statements that refer to common conditions associated with natural states or processes that do not cause significant or permanent harm. • The following are examples of conditions about which structure/function claims could be made under *21 CFR 101.93(g)(2)(iii)* (e.g., morning sickness associated with pregnancy; leg edema associated with pregnancy; mild mood changes, cramps, and edema associated with the menstrual cycle; hot flashes; wrinkles; other signs of aging on the skin, e.g., liver spots, spider veins; presbyopia – inability to change focus from near to far and vice versa associated with aging; mild memory problems associated with aging; hair loss associated with aging; and noncystic acne.) • "Supports a normal healthy attitude during PMS"; "supportive for menopausal women"; and "supports a normal, healthy attitude during PMS" are acceptable structure/function claims.
§101.93(g)(2)(iv): Statement that explicitly or implicitly claims an effect on a disease or diseases through one or more of the following factors: • The name of the product (e.g., "Carpaltum" (carpal tunnel syndrome), "Raynaudin" (Raynaud's phenomenon), "Hepatacure"	Statement that does not explicitly or implicitly claim an effect on disease through one or more of the following factors: • The names of the products that do not imply an effect on a disease (e.g., "Cardiohealth," "Heart Tabs").

Adopted from the Federal Register, Vol. 65, 1/6/00, pp. 999-1050

Chart VI. F.5(e)
(Continued)

Disease Claims (Not Permitted on Dietary Supplement Labels)	Structure/Function Claims (Permitted on Dietary Supplement Labels)
(liver problems). Statements about the formulation of the product, including a claim that the product contains an ingredient (other than an ingredient that is an article included in the definition of "dietary supplement" under 21 U.S.C. 321(ff)(3)) that has been regulated by FDA predominantly as a drug and is well known to consumers for its use in preventing or treating a disease (e.g., aspirin, digoxin, laetrile). • Citation of a publication or reference, if the citation refers to a disease use, and if, in the context of the labeling as a whole, the citation implies treatment or prevention of a disease, e.g., through the placement on the immediate product label or packaging, inappropriate prominence, or lack of relationship to the product's express claims (e.g., labeling for a vitamin E product that included a citation to an article entitled "Serial Coronary Angiographic Evidence That Antioxidant Vitamin Intake Reduces Progression of Coronary Artery Atherosclerosis," would create a disease claim under this criterion; implies treatment or prevention of a disease). • Use of the term "disease" or "diseased," except in general statements about disease prevention that do not refer explicitly or implicitly to a specific disease or class of diseases or to a specific product or ingredient (e.g., "promotes good health and prevents the onset of disease" is a disease claim).	• General statements about health promotion and disease prevention may be acceptable, as long as the statements do not imply that a specific product can diagnose, mitigate, cure, treat, or prevent disease (e.g., "a good diet promotes good health and prevents the onset of disease"). • A picture of a body would not constitute a disease claim. • The term "prescription" or the prescription symbols (Rx), provided that other text does not imply disease.
§101.93(g)(2)(v): • Statement that claims the product belongs to a class of products intended to diagnose, mitigate, treat, cure, or prevent disease (e.g., antibiotic, analgesic, antiviral, diuretic, antimicrobial, antiseptic, antidepressant, vaccine).	• Claims that the product was an "energizer," "rejuvenative," or an "adaptogen," and "laxative" provided the labeling makes clear that the product is not intended to treat chronic constipation.
§101.93(g)(2)(vi): • Statement that explicitly or implicitly claims that the product is a substitute for another product that is a therapy for a disease (e.g., "Herbal Prozac").	• Statement that does not identify a specific drug, drug action, or therapy.

Chart VI. F.5(e)
(Continued)

Disease Claims (Not Permitted on Dietary Supplement Labels)	Structure/Function Claims (Permitted on Dietary Supplement Labels)
§101.93(g)(2)(vii): • Statement that explicitly or implicitly claims that it augments a particular therapy or drug action that is intended to diagnose, mitigate, treat, or prevent a disease or class of diseases ("use as part of your diet when taking insulin to help maintain a healthy blood sugar level").	• Claims that did not identify a specific drug, drug action, or therapy would not constitute a disease claim under this criterion (e.g., "use as a part of your weight loss plan"). • A general statement that a dietary supplement provides nutritional support would be an acceptable structure/function claim, provided that the statement does not suggest that the supplement is intended to augment or have the same purpose as a specific drug, drug action, or therapy for a disease (e.g., "use as part of your diet to help maintain a healthy blood sugar level").
§101.93(g)(2)(viii): • Statement that explicitly or implicitly claims a role in the body's response to a disease or to a vector of disease (a vector of disease is an organism or object that is able to transport or transmit to humans, an agent, such as a virus or bacterium, that is capable of causing disease in man), e.g., "supports the body's antiviral capabilities," "supports the body's ability to resist infection."	• Statement of a more general reference to an effect on a body system that has several functions, only one of which is resistance to disease, would not constitute a disease claim under this criterion (e.g., "supports the immune system" and "vitamin A is necessary to maintaining a healthy immune response").
§101.93(g)(2)(ix):. • Statement that explicitly or implicitly claims to treat, prevent, or mitigate adverse events associated with a therapy for a disease, if the adverse events constitute diseases (e.g., "reduces nausea associated with chemotherapy," "helps avoid diarrhea associated with antibiotic use," "to aid patients with reduced or compromised immune function, such as patients undergoing chemotherapy").	• Statement that does not mention a therapy for disease (e.g., "Helps maintain healthy intestinal flora").
§101.93(g)(2)(x): A statement that would otherwise suggest an effect on a disease or diseases.	

General Guidance

While the context of a claim on the label or in labeling of dietary supplements has to be considered on a case-by-case basis, the following general guidelines are offered as guidance:

1. Statements of nutritional support should provide useful information to consumers about the intended use of a product.

Chart VI. F.5(e)
(Continued)

2. Statements of nutritional support should be supported by scientifically valid evidence substantiating that the statements are truthful and not misleading.

3. Statements indicating the role of a nutrient or dietary ingredient in affecting the structure or function of humans may be made when the statements do not suggest disease prevention or treatment.

Statements that mention a body system, organ, or function affected by the supplement using terms such as "stimulate," "maintain," "support," "regulate," or "promote" can be appropriate when the statements do not suggest disease prevention or treatment or use for health condition that is beyond the ability of the consumer to evaluate.

Section G: Special Labeling for Foods

1. Food Allergen Labeling

(a) Summary Comments

The FDA published a news release that it will require food manufacturers to list food allergens effective January 1, 2006. The Food and Drug Administration (FDA) is requiring food labels to clearly state if food products contain any ingredients that contain protein derived from the eight major allergenic foods. As a result of the Food Allergen Labeling and Consumer Protection Act of 2004 (FALCPA), manufacturers are required to identify in plain English the presence of ingredients that contain protein derived from milk, eggs, fish, crustacean shellfish, tree nuts, peanuts, wheat, or soybeans in the list of ingredients or to say "contains" followed by name of the source of the food allergen after or adjacent to the list of ingredients.

This labeling will be especially helpful to children who must learn to recognize the presence of substances they must avoid. For example, if a product contains the milk-derived protein, casein, the product's label will have to use the term "milk" in addition to the term "casein" so that those with milk allergies can clearly understand the presence of the allergen they need to avoid.

FALCPA does not require food manufacturers or retailers to relabel or remove from grocery or supermarket shelves products that do not reflect the additional allergen labeling as long as the products were labeled before the effective date. As a result, FDA cautions consumers that there will be a transition period of undetermined length during which it is likely that consumers will see packaged food on store shelves and in consumers' homes without the revised allergen labeling. For detailed requirements of FALCPA (see VI.G.1(c)) – Food Allergen Labeling and Consumer Protection Act of 2004 (Title II of Public Law 108-282).

(b) Compliance Policy Guide:

Sec. 555.250 Statement of Policy for Labeling and Preventing Cross-contact of Common Food Allergens[1]

This update to the *Compliance Policy Guides Manual* (August 2000 edition) is a new CPG. This update will be included in the next printing of the *Compliance Policy Guides Manual*. The statements made in the CPG are not intended to create or confer any rights for, or obligations on FDA or any private person, but are intended for internal guidance.

BACKGROUND:

Each year the Food & Drug Administration (FDA) receives reports of consumers who experienced adverse reactions following exposure to an allergenic substance in foods. Food allergies are abnormal responses of the immune system, especially involving the production of allergen specific IgE antibodies, to naturally occurring proteins in certain foods that most individuals can eat safely. Frequently such reactions occur because the presence of the allergenic substances in the foods is not declared on the food label. To combat this problem, the agency issued a letter titled "Notice to Manufacturers," dated June 10, 1996, which addressed labeling issues and Good Manufacturing Practices (GMPs). This letter is available on FDA's Web site, www.cfsan.fda.gov/~lrd/allerg7.html. FDA believes there is scientific consensus that the following foods can cause serious allergic reactions in some individuals and account for more than 90% of all food allergies.[2, 3, 4]

> Peanuts
> Soybeans
> Milk
> Eggs
> Fish
> Crustacea
> Tree nuts
> Wheat

Adopted from FDA' News Release "FDA to Require Food Manufacturers to list Foods Allergens, FDA News, December 210, 2005

Chapter VI

Note: For other foods that may cause an allergic response in certain individuals, the FDA district office should contact CFSAN/Office of *Compliance* for guidance. Manufacturers are responsible for ensuring that food is not adulterated or misbranded as a result of the presence of undeclared allergens. Therefore, the districts should pay particular attention to situations where these substances are added intentionally to food, but not declared on the label, or may be unintentionally introduced into a food product and consequently not declared on the label. When an allergen, not formulated in the product, is identified as likely to occur in the food due to the firm's practices, (e.g., use of common equipment, production scheduling, rework practices) then the district should determine if a manufacturer has identified and implemented control(s) to prevent potential allergen cross-contact, e.g. dedicated equipment, separation, production scheduling, sanitation, proper rework usage (like into like).

POLICY:

Direct addition as ingredients or sub-ingredients

Products which contain an allergenic ingredient by design must comply with *21 U.S.C. 343(i)(2)*. Where substances that are, bear, or contain allergens are added as ingredients or sub-ingredients (including rework), the Federal Food, Drug, and Cosmetic Act (the Act) requires a complete listing of the food ingredients (section 403(i)(2); *21 U.S.C. 343(i)(2); 21 CFR.101.4*) unless a labeling exemption applies.

Exemptions from Ingredient Labeling

Section 403(i)(2) of the Act provides that spices, flavors, and certain colors used in a food may be declared collectively without naming each one. In some instances, these ingredients contain sub-components that are allergens.[5] FDA's regulations *(21 CFR 101.100(a)(3))*, provide that incidental additives, such as processing aids, which are present in a food at insignificant levels and that do not have a technical or functional effect in the finished food are exempt from ingredient declaration. Some manufacturers have asserted to FDA that some allergens that are used as processing aids qualify for this exemption. FDA, however, has never considered food allergens eligible for this exemption. Evidence indicates that some food allergens can cause serious reactions in sensitive individuals upon ingestion of very small amounts; therefore, the presence of an allergen must be declared in accordance with *21 CFR 101.4*. The exemption under *21 CFR 101.100(a)(3)* does not apply to allergenic ingredients.

Practices Used to Prevent Potential Allergen Cross-contact

Allergens may be unintentionally added to food as a result of practices such as improper rework addition, product carry-over due to use of common equipment and production sequencing, or the presence of an allergenic product above exposed product lines. Such practices with respect to allergenic substances may be unsanitary conditions that may render the food injurious to health and adulterate the product under section *402(a)(4)* of the Act [*21 U.S.C. 342(a)(4)]*.

REGULATORY ACTION CRITERIA:

The following represents criteria for direct reference seizure to the Division of Compliance Management and Operations (HFC-210):

1. The FDA district office obtains inspection evidence showing that a food was manufactured to contain an allergenic ingredient as a primary or secondary ingredient, but the food's label does not declare such allergenic ingredient, and
2. The allergenic ingredient is one of the eight (8) ingredients listed in this guide, and
3. The allergenic ingredient was not used as a processing aid in the production of the food, and
4. The inspection of the firm was conducted consistent with the Guide To Inspections of Firms Producing Food Products Susceptible to Contamination with Allergenic Ingredients.

The following represents the criteria for recommending legal action to CFSAN/Office of *
Compliance/Division of Enforcement* (HFS-605):

1. The food contains an undeclared allergenic ingredient that is a derivative of one of the eight (8) ingredients listed in this guide.
2. The food contains an undeclared allergenic ingredient that was used as a processing aid in the manufacture of the product.
3. The food contains an undeclared allergenic ingredient, but the ingredient is not one of the eight (8) allergens listed in this guide.
4. The food is not labeled as containing an allergen, but inspection of the firm shows that it was manufactured under conditions whereby the food may have become contaminated with an allergen.
5. The inspection of the firm was conducted consistent with the Guide To Inspections of Firms Producing Food Products Susceptible to Contamination with Allergenic Ingredients.

Specimen Charges:

Misbranding due to an undeclared allergen:

The article was misbranded when introduced into and while in interstate commerce and is misbranded while held for sale after shipment in interstate commerce, within the meaning of the Act, *21 U.S.C. 343(i)(2)*, in that it is fabricated from two or more ingredients, and its label fails to bear the common or usual name of each such ingredient, namely (specify the undeclared allergenic ingredient).

Adulteration due to food contamination with an allergen:

The article was adulterated when introduced into and while in interstate commerce and is adulterated while held for sale after shipment in interstate commerce, within the meaning of the Act, *21 U.S.C. 342(a)(4),* in that it has been prepared, packed and held under unsanitary conditions whereby it may have been rendered injurious to health.

---

1. This update to the *Compliance Policy Guides Manual* (August 2000 edition) is a new CPG. This update will be included in the next printing of the *Compliance Policy Guides Manual.* The statements made in the CPG are not intended to create or confer any rights for, or obligations on FDA or any private person, but are intended for internal guidance.
2. Food and Agriculture Organization of the United Nations, Report of the FAO Technical Consultation on Food Allergies. Rome, Italy, November 13 to 14, 1995.
3. Hefle, S.L., et al. Allergenic Foods. Critical Reviews in Food Science and Nutrition, 36(S);S69-S89 (1996).
4. Sampson, H.A. Food Allergy, JAMA (278), pp.1888-1894, 1997.
5. As noted in the 1996 letter, FDA is exploring whether allergenic ingredients in spices, flavorings, or colors should be declared, 21 U.S.C. 343(i) notwithstanding. In the meantime, FDA strongly encourages the declaration of an allergenic ingredient of a spice, flavor, or color by either:

   - declaring the allergenic ingredient by it's common or usual name in the ingredient list as a separate ingredient or parenthetically following the term spice, flavor, or color or
   - as a declaration attached at the end of the list of ingredients indicating the presence of a specific allergen.

Material between asterisks is new or revised

Adopted from FDA, CFSAN, Compliance Policy Guides Manual (August 2000 edition), Issued April 19, 2001, Revised May 2005

Chapter VI

(c) **Food Allergen Labeling and Consumer Protection Act of 2004 (Title II of Public Law 108 282)**

SEC. 201. SHORT TITLE.

This title may be cited as the "Food Allergen Labeling and Consumer Protection Act of 2004".

SEC. 202. FINDINGS.

Congress finds that--

  (1) it is estimated that—

    (A) approximately 2 percent of adults and about 5 percent of infants and young children in the United States suffer from food allergies; and

    (B) each year, roughly 30,000 individuals require emergency room treatment and 150 individuals die because of allergic reactions to food;

  (2)

    (A) eight major foods or food groups--milk, eggs, fish, Crustacean shellfish, tree nuts, peanuts, wheat, and soybeans-- account for 90 percent of food allergies;

    (B) at present, there is no cure for food allergies; and

    (C) a food allergic consumer must avoid the food to which the consumer is allergic;

  (3)

    (A) in a review of the foods of randomly selected manufacturers of baked goods, ice cream, and candy in Minnesota and Wisconsin in 1999, the Food and Drug Administration found that 25 percent of sampled foods failed to list peanuts or eggs as ingredients on the food labels; and

    (B) nationally, the number of recalls because of unlabeled allergens rose to 121 in 2000 from about 35 a decade earlier;

  (4) a recent study shows that many parents of children with a food allergy were unable to correctly identify in each of several food labels the ingredients derived from major food allergens;

  (5)

    (A) ingredients in foods must be listed by their "common or usual name";

    (B) in some cases, the common or usual name of an ingredient may be unfamiliar to consumers, and many consumers may not realize the ingredient is derived from, or contains, a major food allergen; and

    (C) in other cases, the ingredients may be declared as a class, including spices, flavorings, and certain colorings, or are exempt from the ingredient labeling requirements, such as incidental additives; and

  (6)

    (A) celiac disease is an immune-mediated disease that causes damage to the gastrointestinal tract, central nervous system, and other organs;

    (B) the current recommended treatment is avoidance of glutens in foods that are associated with celiac disease; and

    (C) a multicenter, multiyear study estimated that the prevalence of celiac disease in the United States is 0.5 to 1 percent of the general population.

SEC. 203. FOOD LABELING; REQUIREMENT OF INFORMATION REGARDING ALLERGENIC SUBSTANCES.

  (a) In General.--Section 403 of the Federal Food, Drug, and Cosmetic Act (21 U.S.C. 343) is amended by adding at the end the following:

    (w)

  (1) If it is not a raw agricultural commodity and it is, or it contains an ingredient that bears or contains, a major food allergen, unless either--

    (A) the word 'Contains', followed by the name of the food source from which the major food allergen is derived, is printed immediately after or is adjacent to the list of ingredients (in a type size no smaller than the type size used in the list of ingredients) required under subsections (g) and (i); or

(B) the common or usual name of the major food allergen in the list of ingredients required under subsections (g) and (i) is followed in parentheses by the name of the food source from which the major food allergen is derived, except that the name of the food source is not required when—

(i) the common or usual name of the ingredient uses the name of the food source from which the major food allergen is derived; or

(ii) the name of the food source from which the major food allergen is derived appears elsewhere in the ingredient list, unless the name of the food source that appears elsewhere in the ingredient list appears as part of the name of a food ingredient that is not a major food allergen under section *201(qq)(2)(A) or(B)*.

(2) As used in this subsection, the term 'name of the food source from which the major food allergen is derived' means the name described in section *201(qq)(1)*; provided that in the case of a tree nut, fish, or Crustacean shellfish, the term 'name of the food source from which the major food allergen is derived' means the name of the specific type of nut or species of fish or Crustacean shellfish.

(3) The information required under this subsection may appear in labeling in lieu of appearing on the label only if the Secretary finds that such other labeling is sufficient to protect the public health. A finding by the Secretary under this paragraph (including any change in an earlier finding under this paragraph) is effective upon publication in the *Federal Register* as a notice.

(4) Notwithstanding subsection (g), (i), or (k), or any other law, a flavoring, coloring, or incidental additive that is, or that bears or contains, a major food allergen shall be subject to the labeling requirements of this subsection.

(5) The Secretary may by regulation modify the requirements of subparagraph (A) or (B) of paragraph (1), or eliminate either the requirement of subparagraph (A) or the requirements of subparagraph (B) of paragraph (1), if the Secretary determines that the modification or elimination of the requirement of subparagraph (A) or the requirements of subparagraph (B) is necessary to protect the public health.

(6)

(A) Any person may petition the Secretary to exempt a food ingredient described in section 201(qq)(2) from the allergen labeling requirements of this subsection.

(B) The Secretary shall approve or deny such petition within 180 days of receipt of the petition or the petition shall be deemed denied, unless an extension of time is mutually agreed upon by the Secretary and the petitioner.

(C) The burden shall be on the petitioner to provide scientific evidence (including the analytical method used to produce the evidence) that demonstrates that such food ingredient, as derived by the method specified in the petition, does not cause an allergic response that poses a risk to human health.

(D) A determination regarding a petition under this paragraph shall constitute final agency action.

(E) The Secretary shall promptly post to a public site all petitions received under this paragraph within 14 days of receipt and the Secretary shall promptly post the Secretary's response to each.

(7)

(A) A person need not file a petition under paragraph (6) to exempt a food ingredient described in section 201(qq)(2) from the allergen labeling requirements of this subsection, if the person files with the Secretary a notification containing

(i) scientific evidence (including the analytical method used) that demonstrates that the food ingredient (as derived by the method specified in the notification, where applicable) does not contain allergenic protein; or

(ii) a determination by the Secretary that the ingredient does not cause an allergic response that poses a risk to human health under a premarket approval or notification program under section 409.

(B) The food ingredient may be introduced or delivered for introduction into interstate commerce as a food ingredient that is not a major food allergen 90 days after the date of receipt of the notification by the Secretary, unless the Secretary determines within the 90-day period that the notification does not meet the requirements of this paragraph, or there is insufficient scientific evidence to determine that the food ingredient does not contain allergenic protein or does not cause an allergenic response that poses a risk to human health.

(C) The Secretary shall promptly post to a public site all notifications received under this subparagraph within 14 days of receipt and promptly post any objections thereto by the Secretary.

(x) Notwithstanding subsection (g), (i), or (k), or any other law, a spice, flavoring, coloring, or incidental additive that is, or that bears or contains, a food allergen (other than a major food allergen), as determined by the Secretary by regulation, shall be disclosed in a manner specified by the Secretary by regulation.

(b) Effect on Other Authority.--The amendments made by this section that require a label or labeling for major food allergens do not alter the authority of the Secretary of Health and Human Services under the Federal Food, Drug, and Cosmetic Act (21 U.S.C. 301 et seq.) to require a label or labeling for other food allergens.

(c) Conforming Amendments.--

(1) Section 201 of the Federal Food, Drug, and Cosmetic Act (21 U.S.C. 321) (as amended by section 102(b)) is amended by adding at the end the following:

(qq) The term 'major food allergen' means any of the following:

(1) Milk, egg, fish (e.g., bass, flounder, or cod), Crustacean shellfish (e.g., crab, lobster, or shrimp), tree nuts (e.g., almonds, pecans, or walnuts), wheat, peanuts, and soybeans.

(2) A food ingredient that contains protein derived from a food specified in paragraph (1), except the following:

(A) Any highly refined oil derived from a food specified in paragraph (1) and any ingredient derived from such highly refined oil.

(B) A food ingredient that is exempt under paragraph (6) or (7) of section 403(w).

(2) Section 403A(a)(2) of the Federal Food, Drug, and Cosmetic Act (21 U.S.C. 343-1(a)(2)) is amended by striking "or 403(i)(2)" and inserting "403(i)(2), 403(w), or 403(x)". (d) Effective Date.--The amendments made by this section shall apply to any food that is labeled on or after January 1, 2006.

SEC. 204. REPORT ON FOOD ALLERGENS.

Not later than 18 months after the date of enactment of this Act, the Secretary of Health and Human Services (in this section referred to as the "Secretary") shall submit to the Committee on Health, Education, Labor, and Pensions of the Senate and the Committee on Energy and Commerce of the House of Representatives a report that--

(1)

(A) analyzes

(i) the ways in which foods, during manufacturing and processing, are unintentionally contaminated with major food allergens, including contamination caused by the use by manufacturers of the same production line to produce both products for which major food allergens are intentional ingredients and products for which major food allergens are not intentional ingredients; and

(ii) the ways in which foods produced on dedicated production lines are unintentionally contaminated with major food allergens; and

(B) estimates how common the practices described in subparagraph (A) are in the food industry, with breakdowns by food type as appropriate;

(2) advises whether good manufacturing practices or other methods can be used to reduce or eliminate cross-contact of foods with the major food allergens;

(3) describes

(A) the various types of advisory labeling (such as labeling that uses the words ``may contain") used by food producers;

(B) the conditions of manufacture of food that are associated with the various types of advisory labeling; and

(C) the extent to which advisory labels are being used on food products;

(4) describes how consumers with food allergies or the caretakers of consumers would prefer that information about the risk of cross-contact be communicated on food labels as determined by using appropriate survey mechanisms;

(5) states the number of inspections of food manufacturing and processing facilities conducted in the previous 2 years and describes--

    (A) the number of facilities and food labels that were found to be in compliance or out of compliance with respect to cross-contact of foods with residues of major food allergens and the proper labeling of major food allergens;

    (B) the nature of the violations found; and

    (C) the number of voluntary recalls, and their classifications, of foods containing undeclared major food allergens; and

(6) assesses the extent to which the Secretary and the food industry have effectively addressed cross-contact issues.

SEC. 205. INSPECTIONS RELATING TO FOOD ALLERGENS.

The Secretary of Health and Human Services shall conduct inspections consistent with the authority under section 704 of the Federal Food, Drug, and Cosmetic Act (21 U.S.C. 374) of facilities in which foods are manufactured, processed, packed, or held--

(1) to ensure that the entities operating the facilities comply with practices to reduce or eliminate cross-contact of a food with residues of major food allergens that are not intentional ingredients of the food; and

(2) to ensure that major food allergens are properly labeled on foods.

SEC. 206. GLUTEN LABELING.

Not later than 2 years after the date of enactment of this Act, the Secretary of Health and Human Services, in consultation with appropriate experts and stakeholders, shall issue a proposed rule to define, and permit use of, the term ``gluten-free'' on the labeling of foods. Not later than 4 years after the date of enactment of this Act, the Secretary shall issue a final rule to define, and permit use of, the term ``gluten-free'' on the labeling of foods.

SEC. 207. IMPROVEMENT AND PUBLICATION OF DATA ON FOOD-RELATED ALLERGIC RESPONSES.

(a) In General.--The Secretary of Health and Human Services, acting through the Director of the Centers for Disease Control and Prevention and in consultation with the Commissioner of Food and Drugs, shall improve (including by educating physicians and other health care providers) the collection of, and publish as it becomes available, national data on--

    (1) the prevalence of food allergies;

    (2) the incidence of clinically significant or serious adverse events related to food allergies; and

    (3) the use of different modes of treatment for and prevention of allergic responses to foods.

(b) Authorization of Appropriations.--For the purpose of carrying out this section, there are authorized to be appropriated such sums as may be necessary.

SEC. 208. FOOD ALLERGIES RESEARCH

(a) In General.--The Secretary of Health and Human Services, acting through the Director of the National Institutes of Health, shall convene an ad hoc panel of nationally recognized experts in allergy and immunology to review current basic and clinical research efforts related to food allergies.

(b) Recommendations.--Not later than 1 year after the date of enactment of this Act, the panel shall make recommendations to the Secretary for enhancing and coordinating research activities concerning food allergies, which the Secretary shall make public.

SEC. 209. FOOD ALLERGENS IN THE FOOD CODE.

The Secretary of Health and Human Services shall, in the Conference for Food Protection, as part of its efforts to encourage cooperative activities between the States under section 311 of the Public Health Service Act (42 U.S.C. 243), pursue revision of the Food Code to provide guidelines for preparing allergen-free foods in food establishments, including in restaurants, grocery store delicatessens and

Adopted from U.S.DHHS, FDA, CFSAN, "Food Allergen Labeling and Consumer Protection Act of 2004", August 2, 2004

bakeries, and elementary and secondary school cafeterias. The Secretary shall consider guidelines and recommendations developed by public and private entities for public and private food establishments for preparing allergen-free foods in pursuing this revision.

SEC. 210. RECOMMENDATIONS REGARDING RESPONDING TO FOOD-RELATED ALLERGIC RESPONSES.

The Secretary of Health and Human Services shall, in providing technical assistance relating to trauma care and emergency medical services to State and local agencies under section 1202(b)(3) of the Public Health Service Act (42 U.S.C. 300d-2(b)(3)), include technical assistance relating to the use of different modes of treatment for and prevention of allergic responses to foods.

(d) Inventory of Notifications Received under 21 U.S.C. 343(w)(7) for Exemptions from Food Allergen Labeling

The Federal Food, Drug, and Cosmetic Act (the Act) governs the labeling of all foods (except meat products, poultry products, and certain egg products, which are regulated by the U.S. Department of Agriculture.) The Food Allergen Labeling and Consumer Protection Act (FALCPA) (Pub L. 108-282) amends the Act's labeling requirements for food ingredients. FALCPAs requirements apply to packaged foods, including conventional foods, dietary supplements, infant formula, and medical foods, all of which are "food" within the Act's definition, 21 U.S.C. 321(f). FALCPA requires that an ingredient (including a flavor, color, or incidental additive) that is a major food allergen, as defined by 21 U.S.C. 321(qq), be more explicitly identified on the food label. Under FALCPA, a "major food allergen" is one of eight foods or food groups (milk, eggs, fish, Crustacean shellfish, tree nuts, wheat, peanuts, and soybeans) or an ingredient that contains protein derived from one of the eight. 21 U.S.C. 321(qq). "Major food allergen" does not include a highly refined oil derived from one of the eight foods or food groups or any ingredient derived from such an oil, as well as any ingredient exempt under a statutory exemption process.

FALCPA establishes a process under 21 U.S.C. 343(w)(7) by which any person may file a notification containing scientific evidence demonstrating that an ingredient does not contain allergenic protein.. The scientific evidence must include the analytical method used and the ingredient must be derived by the specified method. FDA has 90 days to object to a notification. Absent an objection, the food ingredient is exempt from FALCPA's labeling requirements for major food allergens.

FDA is required to post to a public site notifications received under 21 U.S.C. 343(w)(7). This posting is to be made within 14 days of receipt of a notification. The list below reflects the notifications received by FDA that are required at this time to be posted to a public site.

FALN No.	Docket No.	Date Received	Notifier/Ingredient Manufacturer	Description	Major Food Allergen	Agency Response
006	2006FL-0287	July 5, 2006	Nutraceutix, Inc.	lyophilized probiotic cultures	milk	
005	2006FL-120	Mar 3, 2006	Danisco USA	anhydrous lactitol and lactitol monohydrate	milk	
004	2006FL-0017	Dec 27, 2005	Purity Foods, Inc.	Vita Spelt®	wheat	*Objection letter dated 03/27/06*
003	2006FL-0488	Nov 23, 2005	F&A Dairy and International Media and Cultures, Inc.	starter growth media	soy	*Objection letter dated 02/21/06*
002	2006FL-0434	Oct 26, 2005	Ross Products Division, Abbot Laboratories	extensively hydrolyzed casein	milk	*Objection letter dated 01/23/06*
001	2006FL-0416	Sep 30, 2005	Mead Johnson Nutritionals	extensively hydrolyzed casein	milk	*Objection letter dated 12/27/05*

Adopted from U.S.DHHS, FDA, CFSAN, "Food Allergen Labeling and Consumer Protection Act of 2004", August 2, 2004

2. Trans Fat Labeling

(a) Summary Comments

The Food and Drug Administration (FDA) now requires food manufacturers to list *trans* fat (i.e., *trans* fatty acids) on Nutrition Facts and some Supplement Facts panels. Scientific evidence shows that consumption of saturated fat, *trans* fat, and dietary cholesterol raises low-density lipoprotein (LDL or "bad") cholesterol levels that increase the risk of coronary heart disease (CHD). According to the National Heart, Lung, and Blood Institute of the National Institutes of Health, over 12.5 million Americans suffer from CHD, and more than 500,000 die each year. This makes CHD one of the leading causes of death in the United States today.

FDA has required that saturated fat and dietary cholesterol be listed on the food label since 1993. By adding *trans* fat on the Nutrition Facts panel (required by January 1, 2006), consumers now know for the first time how much of all three -- saturated fat, *trans* fat, and cholesterol -- are in the foods they choose. Identifying saturated fat, *trans* fat, and cholesterol on the food label gives consumers information to make heart-healthy food choices that help them reduce their risk of CHD. This revised label, which includes information on *trans* fat as well as saturated fat and cholesterol, will be of particular interest to people concerned about high blood cholesterol and heart disease. However, all Americans should be aware of the risk posed by consuming too much saturated fat, *trans* fat, and cholesterol.

(b) Guidance for Industry and FDA: Requesting an Extension to Use Existing Label Stock after the *Trans* Fat Labeling Effective Date of January 1, 2006

Final Guidance

The purpose of this document is to provide guidance to FDA personnel and the food industry about when and how businesses may request the agency to consider enforcement discretion for the use, on products introduced into interstate commerce on or after the January 1, 2006 effective date, of some or all existing label stock that does not bear *trans* fat labeling in compliance with the *trans* fat final rule. This policy provides guidance to FDA and the food industry related to such requests.

FDA's guidance documents do not establish legally enforceable responsibilities. Instead, guidance documents describe the agency's current thinking on a topic and should be viewed only as recommendations, unless specific regulatory or statutory requirements are cited. The use of the word "should" in agency guidance documents means that something is suggested or recommended, but not required.

FDA published a notice in the *Federal Register* on December 14, 2005 (70 FR 74020) announcing a guidance document about when and how businesses may request the agency to consider enforcement discretion for the use, on products introduced into interstate commerce on or after the January 1, 2006 effective date for the *trans* fat labeling final rule, of some or all existing label stock that does not declare *trans* fat labeling in compliance with the final rule.

FDA received numerous requests from industry in response to this guidance document. Given the agency's limited resources to handle all of these pending requests prior to the January 1, 2006 effective date, the agency intends to consider the exercise of its enforcement discretion for the labeling of foods by firms who have requested an extension postmarked as of December 31, 2005 provided:

1. such requests include all the information outlined in the December 14, 2005 guidance;
2. the labels of the products that are the subject of the request do not bear any statements about *trans* fat;
3. the labels otherwise comply with all FDA labeling requirements, including the Food Allergen Labeling and Consumer Protection Act of 2004 (FALCPA) (Pub. L. 108-282) so that the label accurately reflects ingredients and the nutrient profile, including saturated fat; and
4. the amount of *trans* fat in the product is 0.5 gram or less per labeled serving.

Chapter VI

FDA intends to consider its enforcement discretion for such products subject to a pending request until the agency has responded to the company concerning the request.

This addendum to the guidance does not apply to those who request extensions for labeling postmarked after December 31, 2005 or who have submitted a request for an extension that FDA has denied.

Copies of the entire document are available from the Office of Nutritional Products, Labeling and Dietary Supplements, HFS-800, Center for Food Safety and Applied Nutrition, Food and Drug Administration, 5100 Paint Branch Parkway, College Park, MD 20740, and for questions regarding this guidance, please contact Julie Moss at (301) 436-2373.

Adopted from CFSAN/Office of Nutritional Products, Labeling, and Dietary Supplements, Fat Now Listed With Saturated Fat and Cholesterol on the Nutrition Facts Label January 16, 2004; Updated March 3, 2004; Updated January 1, 2006

(c) Guidance for Industry: Food Labeling, *Trans* Fatty Acids in Nutrition Labeling, Nutrient Content Claims, and Health Claims (Small Entity Compliance Guide)

On July 11, 2003, the Food and Drug Administration (FDA) published a final rule in the *Federal Register* that amended its regulations on food labeling to require that *trans* fatty acids be declared in the nutrition label of conventional foods and dietary supplements (68 FR 41434). This rule is effective January 1, 2006.

FDA's guidance documents, including this guidance, do not establish legally enforceable responsibilities. Instead, guidances describe the Agency's current thinking on a topic and should be viewed only as recommendations, unless specific regulatory or statutory requirements are cited. The use of the word *should* in Agency guidances means that something is suggested or recommended, but not required.

The Food and Drug Administration (FDA) has prepared this Small Entity Compliance Guide in accordance with section 212 of the Small Business Regulatory Fairness Act (P.L. 104-121). This guidance document restates in plain language the legal requirements set forth in *21 CFR 101.9* and *101.36* concerning the declaration of *trans* fatty acids in the nutrition label of conventional foods and dietary supplements, respectively. These regulations are binding and have the force and effect of law. However, this guidance document represents FDA's current thinking on this topic. It does not create or confer any rights for or on any person and does not operate to bind FDA or the public. You can use an alternate approach if the approach satisfies the requirements of the applicable statutes and regulations. If you want to discuss an alternative approach, contact the FDA staff responsible for implementing this guidance. If you cannot identify the appropriate FDA staff, call the appropriate number listed on the title page of this guidance.

This document contains responses to the following questions:

1. What are fatty acids?
2. What are *trans* fatty acids?
3. Why is FDA requiring that *trans* fatty acids be listed in nutrition labeling?
4. Do *trans* fatty acids need to be listed when mono- and polyunsaturated fatty acids are not listed?
5. How should *trans* fatty acids be listed?
6. If a serving contains less than 0.5 gram of *trans* fat, when would "0 g" of *trans* fat not have to be declared?
7. What should be listed as the "% DV" for *trans* fat?
8. Will the content of total fat be changed?
9. Does the final rule change the regulations dealing with nutrient content or health claims?
10. Can *trans* fatty acids be labeled right now?
11. When does this rule become effective?

Adopted from U.S. DHHS, FDA, CFSAN, Food Labeling Trans Fatty Acids in Nutrition Labeling, Nutrient Content Claims, and Health Claims, August 2003

207

This guidance has been prepared by the Division of Nutrition Programs and Labeling, Office of Nutritional Products, Labeling, and Dietary Supplements in the Center for Food Safety and Applied Nutrition (CFSAN) at the U.S. Food and Drug Administration. Copies of the entire report may be obtained from the above mentioned office, and questions regarding this guidance should be directed to Julie Moss at (301) 436-2373.

Chapter VII
Labeling of Food in Special Categories

A. Foods for Special Dietary Uses *(21 CFR Part 105)*

(1) Summary Comments

Section 403(j) of the FD&CA classifies a food as misbranded: "If it purports to be or is represented for special dietary uses, unless its label bears such information concerning its vitamin, mineral, and other dietary properties as the Secretary of Health and Human Services determines to be, and by regulations prescribes as, necessary in order fully to inform purchasers as to its value for such uses."

Section 411(c)(3) of the FD&CA defines "special dietary use" as "a particular use for which a food purports or is represented to be used, including but not limited to the following:

(a) Supplying a special dietary need that exists by reason of a physical, physiological, pathological, or other condition, including but not limited to the conditions of disease, convalescence, pregnancy, lactation, infancy, allergic hypersensitivity to food, underweight, overweight, or the need to control the intake of sodium.

(b) Supplying a vitamin, mineral, or other ingredient for use by humans to supplement the diet by increasing the total dietary intake.

(c) Supplying a special dietary need by reason of being a food for use as the sole item of the diet."

Regulations in *21 CFR Part 105* prescribe appropriate information and statements which must be given on the labels of foods in this class. Importers and foreign shippers should consult the regulations before importing foods represented by labeling or otherwise as foods for special dietary use. When foods for special dietary use are labeled with claims of disease prevention, treatment, mitigation, cure, or diagnosis, they must comply with the drug provisions of the FD&CA.

(2) Questions and Responses

Does the use of the food supply one of the following dietary needs which exists (a) by reason of a physical, physiological, pathological or other condition including, but not limited to, the conditions of disease, convalescence, pregnancy, lactation, allergic hypersensitivity to food, underweight, and overweight or (b) by reason of age including, but not limited to, ages of infancy and childhood:

(a) *Supplementation or fortification of the ordinary or usual diet with any vitamin, mineral, or other dietary property. [Any such particular use of a food is a special dietary use, regardless of whether such food also purports to be or is represented for general use]; or*

(b) *Regulation of the intake of calories and available carbohydrate, or use in the diets of diabetics by use of artificial sweeteners, except when such sweeteners are specifically and solely used for achieving a physical characteristic in the food which cannot be achieved with sugar or other nutritive sweetener?*

[NOTE: For the purposes of these requirements, the terms "infant," "child," and "adult" mean persons not more than 12 months old, more than 12 months but less than 12 years old, and 12 years or more old, respectively.]

    A. If *"YES,"* the product is a special dietary food and, therefore, subject to the requirements of Part 105, continue.
    B. If *"NO,"* STOP HERE, the food is not subject to the requirements of Part 105.

 Does the food that purports to be or is it represented for special dietary use by reason of the decrease or absence of any allergenic property or by reason of being offered as food suitable as a substitute for another food having an allergenic property comply with the following requirements: *(21 CFR 105.62)*

That is, its label bears:

(a). *The common or usual name and the quantity or proportion of each ingredient (including spices, flavoring, and coloring) if the food is fabricated from two or more ingredients;*
(b). *A qualification of the name of the food, or the name of each ingredient thereof if the food is fabricated from two or more ingredients, to reveal clearly the specific plant or animal that is the source of such food or of such ingredient, if such food or such ingredient consists in whole or in part of plant or animal matter and such name does not reveal clearly the specific plant or animal that is such a source; and*
(c) *An informative statement of the nature and effect of any treatment or processing of the food or any ingredient thereof, if the changed allergenic property results from such treatment or processing?*

A. If *"YES,"* continue.
B. If *"NO,"* because the product does not purport to be or is not represented for special dietary use by reason of being a hypoallergenic food or bear labeling in accordance with *21 CFR 105.62*, continue.

    If *"NO,"* because the food purports to be or is represented for special dietary use by reason of being a hypoallergenic food, but its label fails to bear applicable statements required by *21 CFR 105.62*, state that the food _____ special dietary use by reason of _____, but its label fails to bear _____ in accordance with *21 CFR 105.62* (the first blank to be filled with the applicable statement, e.g., purports to be for and/or is represented for; the second blank to be filled with the special dietary usefulness, e.g., a decrease in and/or an absence of lactose in the food; and the third blank to be filled with the specific applicable required label statement omitted from the label, e.g., an informative statement that the lactose has been decreased by use of lactase enzymes).

 Does the food (other than a dietary supplement of vitamins and/or minerals alone) that purports to be or is it represented for special dietary use for infants bear the following labeling as required by 21 CFR 105.65?

(a). *The common or usual name of each ingredient, including spices, flavoring, and coloring, if the food is fabricated from two or more ingredients; and*

(b) If such food or ingredient thereof consists in whole or in part of plant or animal matter and the name of such food or ingredient does not clearly reveal the specific plant or animal which is its source, its name shall be so qualified on the label as to reveal clearly the specific plant or animal that is the source.

A. If *"YES,"* continue.
B. If *"NO,"* because the food does not purport to be or is not represented for special dietary use for infants nor bear labeling statements required by *21 CFR 105.65,* continue.

If *"NO,"* because the food purports to be or is represented for special dietary use for infants, but its label fails to bear applicable label statements required by *21 CFR 105.65,* state that the food _____ special dietary use for infants, but fails to bear _____ on its label in accordance with *21 CFR 105.65* (the first blank to be filled with the statement, e.g., purports to be for and/or is represented for; and the second blank to be filled with the applicable statement(s) required by *21 CFR 105.65,* e.g., the common or usual name of each ingredient, including spices, flavoring, and coloring).

 Does the food that bears label statements relating to its usefulness in reducing or maintaining body weight also bear the following applicable labeling as required by 21 CFR 105.66?
(a). Nutrition labeling in conformity with 21 CFR 101.9, or where applicable, 21 CFR 101.36, unless exempt, and
(b). A conspicuous statement of the basis upon which the food claims to be of special dietary usefulness.

A. If *"YES,"* continue.
B. If *"NO,"* because the food does not bear label statements relating to its usefulness in reducing or maintaining body weight, STOP HERE, the product is not subject to the requirements of *21 CFR 105.66.*

If *"NO,"* because the food bears label statements relating to its usefulness in reducing or maintaining body weight, but fails to bear applicable labeling required by *21 CFR 105.66,* state that the food bears the statement "_____" on its label relating to its usefulness in _____ body weight, but fails to bear _____ as required by _____ (the first blank to be filled with the specific label statement subjecting this product to the requirements of *21 CFR 105.66,* e.g., cheesecake made with nonnutritive sweeteners; the second blank to be filled with statement(s) relating to its usefulness in reducing or maintaining body weight, e.g., maintaining normal weight; the third blank to be filled with the specific required labeling in accordance with *21 CFR 101.9;* and the fourth blank to filled with the specific reference to the regulation violated, e.g., *21 CFR 105.66(a)(1).*

 Does the food that contains a nonnutritive ingredient to achieve its special dietary usefulness (i.e., one not utilized in normal metabolism) bear on its label a statement that it contains a nonnutritive ingredient and the percentage by weight of the nonnutritive ingredient?

A. If *"YES,"* continue.
B. If *"NO,"* because the product contains a nonnutritive ingredient, but the label fails to bear a statement to that effect, state that the product achieves its special dietary usefulness by use of a nonnutritive ingredient, but its label fails to bear a statement that the product contains a nonnutritive ingredient(s) and the percentage by weight of such ingredient(s), as applicable, in accordance with *21 CFR 105.66(b)(1).*

 Does the food that achieves its special dietary usefulness in reducing or maintaining body weight through the use of a nonnutritive sweetener with a nutritive sweetener(s), bear on its label a statement indicating the presence of both types of sweeteners as required by 21 CFR 105.66(b)(2)?

A. If *"YES,"* continue.
B. If *"NO,"* because the food fails to bear a label statement indicating the presence of both sweeteners, state that the food contains nonnutritive and nutritive sweeteners, but its label fails to bear a statement indicating that fact as required by *21 CFR 105.66(b)(2)*.

 Does the special dietary food that purports to be or is represented as a low calorie or reduced calorie food comply with the applicable regulations pertaining to nutrient content claims on calorie content in 21 CFR 101.60(b)(2) and (3), or 21 CFR 101.60(b)(4) and (b)(5) respectively?

A. If *"YES,"* continue.
B. If *"NO,"* because the food does not purport to be or is not represented as a "low calorie" or "reduced calorie" food in its labeling, continue.

If *"NO,"* because the food purports to be or is represented as a low calorie or reduced calorie food, but the label fails to comply with the requirement(s) for these claims, state that the food purports to be and/or is represented as (as appropriate) a _____ but the food fails to comply with the requirement(s) that _____ in accordance with _____ (the first blank to be filled with the specific nutrient content claim, e.g., low calorie food; the second blank to be filled with the specific deviation from the requirement(s), e.g., the food must contain forty calories or less per reference amount customarily consumed; and the third blank to be filled with the applicable CFR reference to the requirement, e.g., *21 CFR 101.60(b)(2)*.

 Is the product a food such as a soft drink that bears terms, such as "diet," "dietetic," "artificially sweetened," or "sweetened with nonnutritive sweetener," that are not false or misleading, that is, labeled as "low calorie" or "reduced calorie" in compliance with 21 CFR 101.60(b)(2) or 21 CFR 101.60(b)(4), respectively, or bears another comparative calorie claim in compliance with other applicable sections of 21 CFR Part 101?

A. If *"YES,"* continue.
B. If *"NO,"* because the product makes no such claims, continue.

If *"NO,"* because the product, such as a soft drink, makes such a claim but fails to comply with the applicable requirement(s), state that the product bears the claim "_____" which suggests that this food is useful as a _____ food which is false and misleading in that the product contains more than _____ calories per reference amount customarily consumed in accordance with the requirement(s) of _____ (the first blank to be filled with the specific claim, e.g., low calorie; the second blank to be filled with a descriptive term, e.g., "diet", the third blank to be filled with the specific deviation from the requirement(s), e.g., 40; and the fourth blank to be filled with the applicable CFR reference to the requirement(s), e.g., *21 CFR 101.60(b)(2)*.

 Does the product that purports to be a special dietary food and bears the term "sugar free" or "no added sugar" comply with 21 CFR 101.60(c) as required by 21 CFR 105.66(f)?

A. If *"YES,"* continue.
B. If *"NO,"* because it does not bear such claims, continue.

If *"NO,"* because it fails to comply with the criteria under *21 CFR 101.60(c),* state that the product label bears the term "_____", but fails to comply with the requirement(s) for the use of the term in that _____ , as required by _____(the first blank to be filled with the specific term used on the label, e.g., sugar free; and the second blank to be filled with the specific deviation from the requirements, e.g., the product provides more than 0.5 g of sugar per reference amount customarily consumed and per labeled serving, and the third blank with the applicable CFR of the requirement).

B. Medical Foods *(21 U.S.C. 360EE(B)(3)—Orphan Drug Act)*

(1) Summary Comments

The term "medical food" was first defined in the Orphan Drug Act Amendments of 1988 [21 USC 360ee (b)(3)] to mean a food which is formulated to be consumed or administered enterally under the supervision of a physician and which is intended for the specific dietary management of a disease or condition for which distinctive nutritional requirements, based on recognized scientific principles, are established by medical evaluation.

The Nutrition Labeling and Education Act of 1990 incorporated, by reference, the definition of medical foods in section 403(q)(5)(A)(iv) of the FD&CA and provided medical foods an exemption from certain nutrition labeling requirements.

FDA incorporated the statutory definition of medical food into the agency's regulations *(21 CFR 101.9(j)(8)).* FDA enumerated criteria intended to clarify the characteristics of a medical food as follows:

(a) It is a specially formulated and processed product (as opposed to naturally occurring foodstuff used in its natural state) for the partial or exclusive feeding of a patient by means of oral intake or enteral feeding by tube;

(b) It is intended for the dietary management of a patient who, because of therapeutic or chronic medical needs, has limited or impaired capacity to ingest, digest, absorb, or metabolize ordinary foodstuffs or certain nutrients, or who has other special medically determined nutrient requirements, the dietary management of which cannot be achieved by the modification of the normal diet alone;

(c) It provides nutritional support specifically modified for the management of the unique nutrient needs that result from the specific disease or condition, as determined by medical evaluation;

(d) It is intended to be used under medical supervision; and

(e) It is intended only for a patient receiving active and ongoing medical supervision wherein the patient requires medical care on a recurring basis for, among other things, instructions on the use of the medical food.

Two sections of the FD&CA contain definitions for foods that are specially formulated and intended to meet the nutritional or dietary requirements of persons whose needs may be different from the healthy general population:

(a) Section 403(q)(5)(A)(iv) defines medical foods, and

(b) Section 411(c)(3) defines foods for special dietary use as food used by man for a particular use for which a food purports or is represented to be used including, but not limited to, the following:

(i) Supplying a special dietary need that exists by reason of a physical, physiological, pathological, or other condition, including but not limited to the condition of disease, convalescence, pregnancy, lactation, infancy, allergic hypersensitivity to food, underweight, overweight, or the need to control the intake of sodium;

(ii) Supplying a vitamin, mineral, or other ingredient for use by man to supplement his diet by increasing the total dietary intake; and

(iii) Supplying a special dietary need by reason of being a food for use as the sole item of the diet.

The following characteristics are differences between medical foods and foods for special dietary use:

(a) The fundamental element of the medical food definition is that the food is intended to meet the distinctive nutritional requirements of a disease or condition that have been established by medical evaluation. FDA interprets this phrase to refer to the body's requirement for specific amounts of nutrients to maintain homeostasis and sustain life; that is the amount of each nutrient that must be available for use in the metabolic and physiological processes necessary to sustain life.

Thus, the distinctive nutritional needs associated with a disease would reflect the total requirement needed by a healthy person, adjusted for the distinctive changes in the nutritional needs of the patient due to the effect of the disease process on metabolism, absorption, or excretion. These distinctive nutritional requirements may be greater than, less than, or in a narrower range of tolerance than for an otherwise healthy individual.

Medical foods, therefore, are formulated to aid in the dietary management of a disease or health-related condition that presents distinctive nutritional requirements that are different from the nutritional requirements of otherwise healthy people.

(b) Foods for special dietary use are foods that are specially formulated to provide only the dietary requirements consistent with recommendations for the general population. Dietary requirements do not reflect a nutritional problem per se; that is, the physiological requirements for nutrients necessary to maintain life or homeostasis addressed by these foods are similar to those of normal, healthy persons.

These foods are formulated in such a way that only the ingredients or the physical form of the diet is different. The quantitative and qualitative nutrient requirements of the person consuming these foods are not significantly different from those of the general population. Examples of persons needing special dietary foods include the following:

- A person who has difficulty swallowing solid food may have a special dietary need for a food that is in liquid form, but this special dietary need doesn't change his physiologic nutrient requirements.

- A person with a gluten-genitive enteropathy requires specially formulated protein in the diet that doesn't contain gluten; however, the actual amount of protein (i.e., amino acids) required by the body is not different from that in otherwise healthy persons.

Thus, foods for special dietary use are intended to meet ordinary nutritional requirements through special dietary means.

C. Infant Formula (Infant Formula Act)

(1) Summary Comments

Infant formulas are liquids or reconstituted powders fed to infants and young children. They serve as substitutes for human milk. Infant formulas have a special role to play in the diets of infants because they are often the only source of nutrients for infants. For this reason, the composition of commercial formulas is carefully controlled and FDA requires that these products meet very strict standards.

Requiring that manufacturers follow specific procedures in manufacturing infant formulas ensures the safety and nutritional quality of infant formulas. In fact, there is a law—known as the Infant Formula Act—which gives FDA special authority to create and enforce standards for commercial infant formulas. Manufacturers must analyze each batch of formula to check nutrient levels and make safety checks. They must also test samples to make sure that the product remains in good condition while it is on the market shelf. In addition, infant formulas must bear codes on their containers to identify each batch; and manufacturers must keep very detailed records of the production and analysis of these foods.

The labeling of infant formulas shall be in compliance with the requirements of *21 CFR Part 107*, Subpart B–Labeling.

D. Dietary Supplements *(Subject is covered in a separate manual.)*

Chapter VIII
Exemptions from FDA Requirements for Foods

Section A: Procedures for Requesting Variations and Exemptions from Required Label Statements *(21 CFR 1.23)*

Section 403(e) of the act (in this part 1, the term "act" means the Federal Food, Drug, and Cosmetic Act) provides for the establishment by regulation of reasonable variations and exemptions for small packages from the required declaration of net quantity of contents. Section 403(i) of the act provides for the establishment by regulation of exemptions from the required declaration of ingredients where such declaration is impracticable, or results in deception or unfair competition. Section 502(b) of the act provides for the establishment by regulation of reasonable variations and exemption for small packages from the required declaration of net quantity of contents. Section 602(b) of the act provides for the establishment by regulation of reasonable variations and exemptions for small packages from the required declaration of net quantity of contents. Section 5(b) of the Fair Packaging and Labeling Act provides for the establishment by regulation of exemptions from certain required declarations of net quantity of contents, identity of commodity, identity and location of manufacturer, packer, or distributor, and from declaration of net quantity of servings represented, based on a finding that full compliance with such required declarations is impracticable or not necessary for the adequate protection of consumers, and a further finding that the nature, form, or quantity of the packaged consumer commodity or other good and sufficient reasons justify such exemptions. The Commissioner, on his own initiative or on petition of an interested person, may propose a variation or exemption based upon any of the foregoing statutory provisions, including proposed findings if section 5(b) of the Fair Packaging and Labeling Act applies, pursuant to parts 10, 12, 13, 14, 15, 16, and 19 of this chapter.

Section B: Exemptions from Required Label Statements *(21 CFR 1.24)*

The following exemptions are granted from label statements required by this part:

1. Foods:
   (a) While held for sale, a food shall be exempt from the required declaration of net quantity of contents specified in this part if said food is received in bulk containers at a retail establishment and is accurately weighed, measured, or counted either within the view of the purchaser or in compliance with the purchaser's order.

(b) Random food packages, as defined in *§101.105(j)* of this chapter, bearing labels declaring net weight, price per pound or per specified number of pounds, and total price shall be exempt from the type size, dual declaration, and placement requirements of *§101.105* of this chapter if the accurate statement of net weight is presented conspicuously on the principal display panel of the package. In the case of food packed in random packages at one place for subsequent shipment and sale at another, the price sections of the label may be left blank provided they are filled in by the seller prior to retail sale. This exemption shall also apply to uniform weight packages of cheese and cheese products labeled in the same manner and by the same type of equipment as random food packages exempted by this paragraph (a)(2) except that the labels shall bear a declaration of price per pound and not price per specified number of pounds.

(c) Individual serving-size packages of foods containing less than ½ ounce or less than ½ fluid ounce for use in restaurants, institutions, and passenger carriers, and not intended for sale at retail, shall be exempt from the required declaration of net quantity of contents specified in this part.

(d) Individually wrapped pieces of penny candy and other confectionery of less than one-half ounce net weight per individual piece shall be exempt from the labeling requirements of this part when the container in which such confectionery is shipped is in conformance with the labeling requirements of this part. Similarly, when such confectionery items are sold in bags or boxes, such items shall be exempt from the labeling requirements of this part, including the required declaration of net quantity of contents specified in this part when the declaration on the bag or box meets the requirements of this part.

(e) (i) Soft drinks packaged in bottles shall be exempt from the placement requirements for the statement of identity prescribed by *§101.3 (a)* and *(d)* of this chapter if such statement appears conspicuously on the bottle closure. When such soft drinks are marketed in a multiunit retail package, the multiunit retail package shall be exempt from the statement of identity declaration requirements prescribed by *§101.3* of this chapter if the statement of identity on the unit container is not obscured by the multiunit retail package.

(ii) A multiunit retail package for soft drinks shall be exempt from the declaration regarding name and place of business required by *§101.5* of this chapter if the package does not obscure the declaration on unit containers or if it bears a statement that the declaration can be found on the unit containers and the declaration on the unit containers complies with *§101.5* of this chapter. The declaration required by *§101.5* of this chapter may appear on the top or side of the closure of bottled soft drinks if the statement is conspicuous and easily legible.

(iii) Soft drinks packaged in bottles which display other required label information only on the closure shall be exempt from the placement requirements for the declaration of contents prescribed by *§101.105(f)* of this chapter if the required content declaration is blown, formed, or molded into the surface of the bottle in close proximity to the closure.

(iv) Where a trademark on a soft drink package also serves as, or is, a statement of identity, the use of such trademark on the package in lines not parallel to the base on which the package rests shall be exempted from the requirement of *§101.3(d)* of this chapter that the statement be in lines parallel to the base so long as there is also at least one statement of identity in lines generally parallel to the base.

(v) A multiunit retail package for soft drinks in cans shall be exempt from the declaration regarding name and place of business required by *§101.5* of this chapter if the package does not obscure the declaration on unit containers or if it bears a statement that the declaration can be found on the unit containers and the declaration on the unit containers complies with *§101.5* of this chapter. The declaration required by *§101.5* of this chapter may appear on the top of soft drinks in cans if the statement is conspicuous and easily legible, provided that when the declaration is embossed, it shall appear in type size at least one-eighth inch in height, or if it is printed, the type size shall not be less than one-sixteenth inch in height. The declaration may follow the curvature of the lid of the can and shall not be removed or obscured by the tab which opens the can.

(f) (i) Ice cream, french ice cream, ice milk, fruit sherbets, water ices, quiescently frozen confections (with or without dairy ingredients), special dietary frozen desserts, and products made in semblance of the foregoing, when measured by and packaged in ½ liquid pint and ½ gallon measure-containers, as defined in the *"Measure Container Code of National*

Bureau of Standards Handbook 44," Specifications, Tolerances, and Other Technical Requirements for Weighing and Measuring Devices, Sec. 4.45 "*Measure-Containers*," which is incorporated by reference, are exempt from the requirements of *§101.105(b)(2)* of this chapter to the extent that net contents of 8-fluid ounces and 64-fluid ounces (or 2 quarts) may be expressed as ½ pint and ½ gallon, respectively. Copies are available from the Division of Regulatory Guidance, Center for Food Safety and Applied Nutrition (HFF-310), Food and Drug Administration, 200 C St. SW., Washington, DC 20204, or available for inspection at the Office of the *Federal Register*, 800 North Capitol Street NW., suite 700, Washington, DC.

    (ii) The foods named in paragraph (a)(6)(i) of this section, when measured by and packaged in 1-liquid pint, 1-liquid quart, and ½ gallon measure-containers, as defined in the "*Measure Container Code of National Bureau of Standards Handbook 44*," Specifications, Tolerances, and Other Technical Requirements for Weighing and Measuring Devices, Sec. 4.45 "*Measure-Containers*," which is incorporated by reference, are exempt from the dual net contents declaration requirement of *§101.105(j)* of this chapter. Copies are available from the Division of Regulatory Guidance, Center for Food Safety and Applied Nutrition (HFF-310), Food and Drug Administration, 200 C St. SW., Washington, DC 20204, or available for inspection at the Office of the Federal Register, 800 North Capitol Street NW., suite 700, Washington, DC.

    (iii) The foods named in paragraph (a)(6)(i) of this section, when measured by and packaged in ½-liquid pint, 1-liquid pint, 1-liquid quart, ½-gallon, and 1-gallon measured-containers, as defined in the "*Measure Container Code of National Bureau of Standards Handbook 44*," Specifications, Tolerances, and Other Technical Requirements for Weighing and Measuring Devices, Sec. 4.45 "*Measure-Containers*," which is incorporated by reference, are exempt from the requirement of *§101.105(f)* of this chapter that the declaration of net contents be located within the bottom 30 percent of the principal display panel. Copies are available from the Division of Regulatory Guidance, Center for Food Safety and Applied Nutrition (HFF-310), Food and Drug Administration, 200 C St. SW., Washington, DC 20204, or available for inspection at the Office of the Federal Register, 800 North Capitol Street NW., suite 700, Washington, DC.

(g) (i) Milk, cream, light cream, coffee or table cream, whipping cream, light whipping cream, heavy or heavy whipping cream, sour or cultured sour cream, half-and-half, sour or cultured half-and-half, reconstituted or recombined milk and milk products, concentrated milk and milk products, skim or skimmed milk, vitamin D milk and milk products, fortified milk and milk products, homogenized milk, flavored milk and milk products, buttermilk, cultured buttermilk, cultured milk or cultured whole buttermilk, low-fat milk (0.5 to 2.0 percent butterfat), and acidified milk and milk products, when packaged in containers of 8- and 64-fluid-ounce capacity, are exempt from the requirements of *§101.105(b)(2)* of this chapter to the extent that net contents of 8 fluid ounces and 64 fluid ounces (or 2 quarts) may be expressed as ½ pint and ½ gallon, respectively.

    (ii) The products listed in paragraph (a)(7)(i) of this section, when packaged in glass or plastic containers of ½-pint, 1-pint, 1-quart, ½-gallon, and 1-gallon capacities are exempt from the placement requirement of *§101.105(f)* of this chapter that the declaration of net contents be located within the bottom 30 percent of the principal display panel provided that other required label information is conspicuously displayed on the cap or outside closure and the required net quantity of contents declaration is conspicuously blown, formed, or molded into or permanently applied to that part of the glass or plastic container that is at or above the shoulder of the container.

    (iii) The products listed in paragraph (a)(7)(i) of this section, when packaged in containers of 1-pint, 1-quart, and ½-gallon capacities are exempt from the dual net-contents declaration requirement of *§ 101.105(j)* of this chapter.

(h) Wheat flour products, as defined by *§§ 137.105, 137.155, 137.160, 137.165, 137.170, 137.175, 137.180, 137.185, 137.200,* and *137.205* of this chapter, packaged:

    (i) In conventional 2-, 5-, 10-, 25-, 50- and 100-pound packages are exempt from the placement requirement of *§101.105(f)* of this chapter that the declaration of net contents be located within the bottom 30 percent of the area of the principal display panel of the label; and

(ii) In conventional 2-pound packages are exempt from the dual net-contents declaration requirement of *§101.105(j)* of this chapter provided the quantity of contents is expressed in pounds.

(i) (i) Twelve shell eggs packaged in a carton designed to hold 1 dozen eggs and designed to permit the division of such carton by the retail customer at the place of purchase into two portions of one-half dozen eggs each are exempt from the labeling requirements of this part with respect to each portion of such divided carton if the carton, when undivided, is in conformance with the labeling requirements of this part.

(ii) Twelve shell eggs packaged in a carton designed to hold 1 dozen eggs are exempt from the placement requirements for the declaration of contents prescribed by *§101.105(f)* of this chapter if the required content declaration is otherwise placed on the principal display panel of such carton and if, in the case of such cartons designed to permit division by retail customers into two portions of one-half dozen eggs each, the required content declaration is placed on the principal display panel in such a manner that the context of the content declaration is destroyed upon division of the carton.

(j) Butter as defined in 42 Stat. 1500 (excluding whipped butter):

(i) In 8-ounce and in 1-pound packages is exempt from the requirements of *§101.105(f)* of this chapter that the net contents declaration be placed within the bottom 30 percent of the area of the principal display panel;

(ii) In 1-pound packages is exempt from the requirements of *§101.105(j)(1)* of this chapter that such declaration be in terms of ounces and pounds, to permit declaration of "*I-pound*" or "*one pound*"; and

(iii) In 4-ounce, 8-ounce, and 1-pound packages with continuous label copy wrapping is exempt from the requirements of *§§101.3* and *101.105(f)* of this chapter that the statement of identity and net contents declaration appear in lines generally parallel to the base on which the package rests as it is designed to be displayed, provided that such statement and declaration are not so positioned on the label as to be misleading or difficult to read as the package is customarily displayed at retail.

(k) Margarine as defined in *§166.110* of this chapter and imitations thereof in 1-pound rectangular packages, except for packages containing whipped or soft margarine or packages that contain more than four sticks, are exempt from the requirement of *§101.105(f)* of this chapter that the declaration of the net quantity of contents appear within the bottom 30 percent of the principal display panel and from the requirement of *§101.105(j)(1)* of this chapter that such declaration be expressed both in ounces and in pounds to permit declaration of "1-pound" or "one pound," provided an accurate statement of net weight appears conspicuously on the principal display panel of the package.

(l) Corn flour and related products, as they are defined by *§§137.211, 137.215*, and *§§ 137.250* through *137.290* of this chapter, packaged in conventional 5-, 10-, 25-, 50-, and 100-pound bags are exempt from the placement requirement of *§101.105(f)* of this chapter that the declaration of net contents be located within the bottom 30 percent of the area of the principal display panel of the label.

(m) (i) Single strength and less than single strength fruit juice beverages, imitations thereof, and drinking water when packaged in glass or plastic containers of ½-pint, 1-pint, 1-quart, ½-gallon, and 1-gallon capacities are exempt from the placement requirement of *§101.105(f)* of this chapter that the declaration of net contents be located within the bottom 30 percent of the principal display panel: Provided, That other required label information is conspicuously displayed on the cap or outside closure and the required net quantity of contents declaration is conspicuously blown, formed, or molded into or permanently applied to that part of the glass or plastic container that is at or above the shoulder of the container.

(ii) Single strength and less than single strength fruit juice beverages, imitations thereof, and drinking water when packaged in glass, plastic, or paper (fluid milk type) containers of 1-pint, 1-quart, and ½-gallon capacities are exempt from the dual net-contents declaration requirement of *§101.105(j)* of this chapter.

(iii) Single strength and less than single strength fruit juice beverages, imitations thereof, and drinking water when packaged in glass, plastic, or paper (fluid milk type) containers of 8- and 64-fluid-ounce capacity, are exempt from the requirements *of §101.105(b)(2)* of this chapter to the extent that net contents of 8 fluid ounces and 64 fluid ounces (or 2 quarts) may be expressed as ½ pint (or half pint) and ½ gallon (or half gallon), respectively.

(n) The unit containers in a multiunit or multicomponent retail food package shall be exempt from regulations of section 403 (e)(1), (g)(2), (i)(2), (k), and (q) of the act with respect to the requirements for label declaration of the name and place of business of the manufacturer, packer, or distributor; label declaration of ingredients; and nutrition information when:

(i) The multiunit or multicomponent retail food package labeling meets all the requirements of this part;

(ii) The unit containers are securely enclosed within and not intended to be separated from the retail package under conditions of retail sale; and

(iii) Each unit container is labeled with the statement "*This Unit Not Labeled For Retail Sale*" in type size not less than one-sixteenth of an inch in height. The word "*Individual*" may be used in lieu of or immediately preceding the word "*Retail*" in the statement.

Section C: Petitions Requesting Exemption from Preemption for State or Local Requirements
(21 CFR 100.1)

1   Scope and purpose:

(a) This subpart applies to the submission and consideration of petitions under section 403A(b) of the Federal Food, Drug, and Cosmetic Act (the act), by a State or a political subdivision of a State, requesting exemption of a State requirement from preemption under section 403A(a) of the act.

(b) Section 403A(b) of the act provides that where a State requirement has been preempted under section 403A(a) of the act, the State may petition the agency for an exemption. The agency may grant the exemption, under such conditions as it may prescribe by regulation, if the agency finds that the State requirement will not cause any food to be in violation of any applicable requirement under Federal law, will not unduly burden interstate commerce, and is designed to address a particular need for information that is not met by the preemptive Federal requirement.

2. Definitions:

(a) *Act* means the Federal Food, Drug, and Cosmetic Act (21 U.S.C. 321 et seq.).

(b) *Agency* means the Food and Drug Administration.

(c) *Commissioner* means the Commissioner of Food and Drugs.

(d) *State* means a State as defined in section 201(a)(1) of the act (which includes a territory of the United States, the District of Columbia, and Puerto Rico) or any political subdivision of a State having authority to issue food standards and food labeling regulations having force of law.

(e) *State requirement* means any statute, standard, regulation, or other requirement that is issued by a State.

3. Prerequisites for petitions for exemption from preemption:

The Food and Drug Administration will consider a petition for exemption from preemption on its merits only if the petition demonstrates that:

(a) The State requirement was enacted or was issued as a final rule by an authorized official of the State and is in effect or would be in effect but for the provisions of section 403A of the act.

(b) The State requirement is subject to preemption under section 403A(a) of the act because of a statutory provision listed in that section or because of a Federal standard or other Federal regulation that is in effect, or that has been published as a final rule with a designated effective date, and that was issued under the authority of a statutory provision listed in that section. For the purposes of this subpart, all petitions seeking exemption from preemption under section 403A(a)(3) through (a)(5) of the act submitted before May 8, 1992, will be considered timely even though the applicable statutory provisions or regulations are not yet in effect.

(c) The petitioner is an official of a State having authority to act for, or on behalf of, the Government in applying for an exemption of State requirements from preemption.

(d) The State requirement is subject to preemption under section 403A(a) of the act because it is not identical to the requirement of the preemptive Federal statutory provision or regulation including a standard of identity, quality, and fill. "*Not identical to*" does not refer to the specific words in the requirement but instead means that the State requirement directly or indirectly imposes obligations or contains provisions concerning the composition or labeling of food, or concerning a

food container, that:
  (i) Are not imposed by or contained in the applicable provision (including any implementing regulation) of section 401 or 403 of the act; or
  (ii) Differ from those specifically imposed by or contained in the applicable provision (including any implementing regulation) of section 401 or 403 of the act.
4. Form of petition:
  (a) All information included in the petition should meet the general requirements of *§10.20(c)* of this chapter.
  (b) An original and one copy of the petition shall be submitted, or the petitioner may submit an original and a computer readable disk containing the petition. Contents of the disk should be in a standard format, such as ASCII format. (Petitioners interested in submitting a disk should contact the Center for Food Safety and Applied Nutrition for details.)
  (c) Petitions for exemption from preemption for a State requirement shall be submitted to the Dockets Management Branch in the following form:
  (Date)_____
  Dockets Management Branch,
  Food and Drug Administration,
  Department of Health and Human
  Services,
  Rm. 1-23,12420 Parklawn Dr.,
  Rockville, MD 20857.

PETITION REQUESTING EXEMPTION FROM PREEMPTION FOR STATE REQUIREMENT

The undersigned submits this petition under section 403A(b) of the Federal Food, Drug, and Cosmetic Act to request that the Food and Drug Administration exempt a State requirement from preemption. The undersigned has authority to act for, or on behalf of, the (identify State or political subdivision of the State) because (document petitioner's authority to submit petition on behalf of the State).

A. Action Requested

1. Identify and give the exact wording of the State requirement and give date it was enacted or issued in final form.
2. Identify the specific standard or regulation that is believed to preempt the State requirement and the section and paragraph of the act that the standard or regulation implements.

B. Documentation of State Requirement

Provide a copy of the State requirement that is the subject of the application. Where available, the application should also include copies of any legislative history or background materials used in issuing the requirement, including hearing reports or studies concerning the development or consideration of the requirement.

C. Statement of Grounds

A petition for an exemption from preemption should contain the following:

1. An explanation of the State requirement and its rationale, and a comparison of State and Federal requirements to show differences.
2. An explanation of why compliance with the State requirement would not cause a food to be in violation of any applicable requirement under Federal law.
3. Information on the effect that granting the State petition will have on interstate commerce. The petition should contain information on economic feasibility, i.e., whether the State and Federal requirements have significantly different effects on the production and distribution of the food product; comparison of the costs of compliance as shown by data or information on the actual or anticipated effect of the State and Federal requirements on the sale and price of the food product in interstate commerce; and the effect of the State requirement on the availability of the food product to consumers. To the extent possible, the petition should include information showing that it is practical and feasible for producers of food products to comply with the State requirement. Such information may be submitted in the form of statements from affected persons indicating their ability to comply.
4. Identification of a particular need for information that the State requirement is designed to meet, which need is not met by Federal law. The petition should describe the conditions that require the State to petition for an exemption, the information need that the State requirement fulfills, the inadequacy of the Federal requirement in addressing this need, and the geographical area or political subdivision in which such need exists.

D. Environmental Impact

The petition shall contain a claim for categorical exclusion under *21 CFR 25.24* or an environmental assessment under *2l CFR 25.31*.

E. Notification

Provide name and address of person, branch, department, or other instrumentality of the State government that should be notified of the Commissioner's action concerning the petition.

F. Certification

The undersigned certifies that, to the best knowledge and belief of the undersigned, this petition includes all information and views on which the petition relies.

(Signature)----------------------
(Name of petitioner)-------------
(Mailing address)---------------
(Telephone number)-------------
(Information collection requirements in this section were approved by the Office of Management and Budget (OMB) and assigned OMB number 0910-0277)

5. Submission of petition for exemption; public disclosure:
   The availability for public disclosure of a petition for exemption will be governed by the rules specified in §10.20(j) of this chapter.
6. Agency consideration of petitions:
   (1) Unless otherwise specified in this section, all relevant provisions and requirements of subpart B of part 10 of this chapter, are applicable to State petitions requesting exemption from Federal preemption under section 403A(b) of the act.
   (2) If a petition does not meet the prerequisite requirements of paragraph (c) of this section, the agency will issue a letter to the petitioner denying the petition and stating in what respect the petition does not meet these requirements.

(3) If a petition appears to meet the prerequisite requirements in paragraph (c) of this section, it will be filed by the Dockets Management Branch, stamped with the date of filing, and assigned a docket number. The docket number identifies the file established by the Dockets Management Branch for all submissions relating to the petition, as provided in this part. Subsequent submissions relating to the matter must refer to the docket number and will be filed in the docket file. The Dockets Management Branch will promptly notify the petitioner in writing of the filing and docket number of a petition.

(4) Any interested person may submit written comments to the Dockets Management Branch on a filed petition as provided in *§10.30(d)* of this chapter.

(5) Within 90 days of the date of filing the agency will furnish a response to the petitioner. The response will either:

(i) State that the agency has tentatively determined that the petition merits the granting of an exemption, and that it intends to publish in the *FEDERAL REGISTER* a proposal to grant the exemption through rulemaking;

(ii) Deny the petition and state the reasons for such denial; or

(iii) Provide a tentative response indicating why the agency has been unable to reach a decision on the petition, e.g., because of other agency priorities or a need for additional information.

7. If a State submitted a petition for exemption of a State requirement from preemption under section 403A(a)(3) through (a)(5) of the act before May 8, 1992, that State requirement will not be subject to preemption until:

(1) November 8, 1992, or

(2) Action on the petition, whichever occurs later.

Section D: Food: Exemptions from Labeling (*21 CFR 101.100*)

a) The following foods are exempt from compliance with the requirements of section 403(i)(2) of the act (requiring a declaration on the label of the common or usual name of each ingredient when the food is fabricated from two or more ingredients).

(1) An assortment of different items of food, when variations in the items that make up different packages packed from such assortment normally occur in good packing practice and when such variations result in variations in the ingredients in different packages, with respect to any ingredient that is not common to all packages. Such exemption, however, shall be on the condition that the label shall bear, in conjunction with the names of such ingredients as are common to all packages, a statement (in terms that are as informative as practicable and that are not misleading) indicating by name other ingredients which may be present.

(2) A food having been received in bulk containers at a retail establishment, if displayed to the purchaser with either:

(i) The labeling of the bulk container plainly in view, provided ingredient information appears prominently and conspicuously in lettering of not less than one-fourth of an inch in height; or

(ii) A counter card, sign, or other appropriate device bearing prominently and conspicuously, but in no case with lettering of less than one-fourth of an inch in height, the information required to be stated on the label pursuant to section 403(i)(2) of the Federal Food, Drug, and Cosmetic Act (the act).

(3) Incidental additives that are present in a food at insignificant levels and do not have any technical or functional effect in that food. For the purposes of this paragraph (a)(3), incidental additives are:

(i) Substances that have no technical or functional effect but are present in a food by reason of having been incorporated into the food as an ingredient of another food, in which the substance did have a functional or technical effect.

(ii) Processing aids, which are as follows:

(a) Substances that are added to a food during the processing of such food but are removed in some manner from the food before it is packaged in its finished form.

(b) Substances that are added to a food during processing, are converted into constituents normally present in the food, and do not significantly increase the amount of the constituents naturally found in the food.

(c) Substances that are added to a food for their technical or functional effect in the processing but are present in the finished food at insignificant levels and do not have any technical or functional effect in that food.

(iii) Substances migrating to food from equipment or packaging or otherwise affecting food that are not food additives as defined in section 201(s) of the act; or if they are food additives as so defined, they are used in conformity with regulations established pursuant to section 409 of the act.

(4) For the purposes of paragraph (a)(3) of this section, any sulfiting agent (sulfur dioxide, sodium sulfite, sodium bisulfite, potassium bisulfite, sodium metabisulfite, and potassium metabisulfite) that has been added to any food or to any ingredient in any food and that has no technical effect in that food will be considered to be present in an insignificant amount only if no detectable amount of the agent is present in the finished food. A detectable amount of sulfiting agent is 10 parts per million or more of the sulfite in the finished food. Compliance with this paragraph will be determined using sections 20.123-20.125, *"Total Sulfurous Acid,"* in *"Official Methods of Analysis of the Association of Official Analytical Chemists,"* 14th Ed. (1984), which is incorporated by reference and the refinements of the *"Total Sulfurous Acid"* procedure in the *"Monier-Williams Procedure (with Modifications) for Sulfites in Foods,"* which is Appendix A to Part 101. A copy of sections 20.123-20.125 of the *"Official Methods of Analysis of the Association of Official Analytical Chemists"* is available from the Association of Official Analytical Chemists, P.O. Box 540, Benjamin Franklin Station, Washington, DC 20044, or available for inspection at the Office of the Federal Register, 800 North Capitol Street, NW., suite 700, Washington, DC.

b) A food repackaged in a retail establishment is exempt from the following provisions of the act if the conditions specified are met.

(1) Section 403(e)(1) of the act (requiring a statement on the label of the name and place of business of the manufacturer, packer, or distributor).

(2) Section 403(g)(2) of the act (requiring the label of a food which purports to be or is represented as one for which a definition and standard of identity has been prescribed to bear the name of the food specified in the definition and standard and, insofar as may be required by the regulation establishing the standard using the common names of the optional ingredients present in the food), if the food is displayed to the purchaser with its interstate labeling clearly in view, or with a counter card, sign, or other appropriate device bearing prominently and conspicuously the information required by these provisions.

(3) Section 403(i)(1) of the act (requiring the label to bear the common or usual name of the food), if the food is displayed to the purchaser with its interstate labeling clearly in view, or with a counter card, sign, or other appropriate device bearing prominently and conspicuously the common or usual name of the food, or if the common or usual name of the food is clearly revealed by its appearance.

c) An open container (a container of rigid or semirigid construction, which is not closed by lid, wrapper, or otherwise other than by an uncolored transparent wrapper which does not obscure the contents) of a fresh fruit or fresh vegetable, the quantity of contents of which is not more than 1 dry quart, shall be exempt from the labeling requirements of sections 403(e), (g)(2) (with respect to the name of the food specified in the definition and standard), and (i)(l) of the act; but such exemption shall be on the condition that if two or more such containers are enclosed in a crate or other shipping package, such crate or package shall bear labeling showing the number of such containers enclosed therein and the quantity of the contents of each.

d) Except as provided by paragraphs (e) and (f) of this section, a shipment or other delivery of a food which is, in accordance with the practice of the trade, to be processed, labeled, or repacked in substantial quantity at an establishment other than that where originally processed or packed, shall be exempt, during the time of introduction into and movement in interstate commerce and the time of holding in such establishment, from compliance with the labeling requirements of section 403 (c), (e), (g), (h), (i), (k), and (q) of the act if:

(1) The person who introduced such shipment or delivery into interstate commerce is the operator of the establishment where such food is to be processed, labeled, or repacked; or

(2) In case such person is not such operator, such shipment or delivery is made to such establishment under a written agreement, signed by and containing the post office addresses of such person and such operator, and containing such specifications for the processing, labeling,

or repacking, as the case may be, of such food in such establishment as will ensure, if such specifications are followed, that such food will not be adulterated or misbranded within the meaning of the act upon completion of such processing, labeling, or repacking. Such person and such operator shall each keep a copy of such agreement until 2 years after the final shipment or delivery of such food from such establishment, and shall make such copies available for inspection at any reasonable hour to any officer or employee of the Department who requests them.

(3) The article is an egg product subject to a standard of identity promulgated in part 160 of this chapter, is to be shipped under the conditions specified in paragraph (d) (1) or (2) of this section and for the purpose of pasteurization or other treatment as required in such standard, and each container of such egg product bears a conspicuous tag or label reading "*Caution—This egg product has not been pasteurized or otherwise treated to destroy viable Salmonella microorganisms*". In addition to safe and suitable bactericidal processes designed specifically for Salmonella destruction in egg products, the term "other treatment" in the first sentence of this paragraph shall include use in acidic dressings in the processing of which the pH is not above 4.1 and the acidity of the aqueous phase, expressed as acetic acid, is not less than 1.4 percent, subject also to the conditions that:

(i) The agreement required in paragraph (d) (2) of this section shall also state that the operator agrees to utilize such unpasteurized egg products in the processing of acidic dressings according to the specifications for pH and acidity set forth in this paragraph, agrees not to deliver the acidic dressing to a user until at least 72 hours after such egg product is incorporated in such acidic dressing, and agrees to maintain for inspection adequate records covering such processing for 2 years after such processing.

(ii) In addition to the caution statement referred to above, the container of such egg product shall also bear the statement "*Unpasteurized _____ for use in acidic dressings only*", the blank being filled in with the applicable name of the eggs or egg product.

e) Conditions affecting expiration of exemptions:

(1) An exemption of a shipment or other delivery of a food under paragraph (d) (1) or (3) of this section shall, at the beginning of the act of removing such shipment or delivery, or any part thereof, from such establishment become void ab initio if the food comprising such shipment, delivery, or part is adulterated or misbranded within the meaning of the act when so removed.

(2) An exemption of a shipment or other delivery of a food under paragraph (d) (2) or (3) of this section shall become void ab initio with respect to the person who introduced such shipment or delivery into interstate commerce upon refusal by such person to make available for inspection a copy of the agreement, as required by paragraph (d) (2) or (3) of this section.

(3) An exemption of a shipment or other delivery of a food under paragraph (d) (2) or (3) of this section shall expire:

(i) At the beginning of the act of removing such shipment or delivery, or any part thereof, from such establishment if the food constituting such shipment, delivery, or part is adulterated or misbranded within the meaning of the act when so removed; or

(ii) Upon refusal by the operator of the establishment where such food is to be processed, labeled, or repacked, to make available for inspection a copy of the agreement, as required by such paragraph.

f) The word "*processed*" as used in this paragraph shall include the holding of cheese in a suitable warehouse at a temperature of not less than 35° F for the purpose of aging or curing to bring the cheese into compliance with requirements of an applicable definition and standard of identity. The exemptions provided for in paragraph (d) of this section shall apply to cheese which is, in accordance with the practice of the trade, shipped to a warehouse for aging or curing, on condition that the cheese is identified in the manner set forth in one of the applicable following paragraphs, and in such case the provisions of paragraph (e) of this section shall also apply:

(1) In the case of varieties of cheese for which definitions and standards of identity require a period of aging whether or not they are made from pasteurized milk, each such cheese shall bear on the cheese a legible mark showing the date at which the preliminary manufacturing process has been completed and at which date curing commences, and to each cheese, on its wrapper or immediate container, shall be affixed a removable tag bearing the statement "*Uncured _____ cheese for completion of curing and proper labeling*", the blank being filled in with the applicable name of the variety of cheese. In the case of Swiss cheese, the date at which the preliminary

manufacturing process had been completed and at which date curing commences is the date on which the shaped curd is removed from immersion in saturated salt solution as provided in the definition and standard of identity for Swiss cheese, and such cheese shall bear a removable tag reading, "*To be cured and labeled as 'Swiss cheese,' but if eyes do not form, to be labeled as 'Swiss cheese for manufacturing'*".

(2) In the case of varieties of cheeses which when made from unpasteurized milk are required to be aged for not less than 60 days, each such cheese shall bear a legible mark on the cheese showing the date at which the preliminary manufacturing process has been completed and at which date curing commences, and to each such cheese or its wrapper or immediate container shall be affixed a removable tag reading, "_____*cheese made from unpasteurized milk. For completion of curing and proper labeling*", the blank being filled in with the applicable name of the variety of cheese.

(3) In the case of cheddar cheese, washed curd cheese, colby cheese, granular cheese, and brick cheese made from unpasteurized milk, each such cheese shall bear a legible mark on the cheese showing the date at which the preliminary manufacturing process has been completed and at which date curing commences, and to each such cheese or its wrapper or immediate container shall be affixed a removable tag reading "_____ *cheese made from unpasteurized milk. For completion of curing and proper labeling, or for labeling as* _____ *cheese for manufacturing*", the blanks being filled in with the applicable name of the variety of cheese.

g) The label declaration of a harmless marker used to identify a particular manufacturer's product may result in unfair competition through revealing a trade secret. Exemption from the label declaration of such a marker is granted, therefore, provided that the following conditions are met:

(1) The person desiring to use the marker without label declaration of its presence has submitted to the Commissioner of Food and Drugs full information concerning the proposed usage and the reasons why he believes label declaration of the marker should be subject to this exemption; and

(2) The person requesting the exemption has received from the Commissioner of Food and Drugs a finding that the marker is harmless and that the exemption has been granted.

h) Wrapped fish fillets of nonuniform Weight intended to be unpacked and marked with the correct weight at or before the point of retail sale in an establishment other than that where originally packed shall be exempt from the requirement of section 403(e)(2) of the act during introduction and movement in interstate commerce and while held for sale prior to weighing and marking:

(1) *Provided*, That
    (i) The outside container bears a label declaration of the total net weight; and
    (ii) The individual packages bear a conspicuous statement "*To be weighed at or before time of sale*" and a correct statement setting forth the weight of the wrapper;

(2) *Provided further*, That it is the practice of the retail establishment to weigh and mark the individual packages with a correct net-weight statement prior to or at the point of retail sale. A statement of the weight of the wrapper shall be set forth so as to be readily read and understood, using such term as "*wrapper tare _____ ounce*", the blank being filled in with the correct average weight of the wrapper used.

(3) The act of delivering the wrapped fish fillets during the retail sale without the correct net-weight statement shall be deemed an act that results in the product's being misbranded while held for sale. Nothing in this paragraph shall be construed as requiring net weight statements for wrapped fish fillets delivered into institutional trade provided the outside container bears the required information.

i) Wrapped clusters (consumer units) of bananas of nonuniform weight intended to be unpacked from a master carton or container and weighed at or before the point of retail sale in an establishment other than that where originally packed shall be exempt from the requirements of section 403(e)(2) of the act during introduction and movement in interstate commerce and while held for sale prior to weighing:

(1) *Provided*, That
    (i) The master carton or container bears a label declaration of the total net weight; and
    (ii) The individual packages bear a conspicuous statement "*To be weighed at or before the time of sale*" and a correct statement setting forth the weight of the wrapper; using such term as "*wrapper tare_____ounce*", the blank being filled in with the correct average weight of the wrapper used;

(2) *Provided further*, That it is the practice of the retail establishment to weigh the individual packages either prior to or at the time of retail sale.

(3) The act of delivering the wrapped clusters (consumer units) during the retail sale without an accurate net weight statement or alternatively without weighing at the time of sale shall be deemed an act which results in the product's being misbranded while held for sale. Nothing in this paragraph shall be construed as requiring net-weight statements for clusters (consumer units) delivered into institutional trade, provided that the master container or carton bears the required information.

Section E: Temporary Exemption for Purposes of Conducting Authorized Food Labeling Experiments *(21 CFR 101.108)*

a) The food industry is encouraged to experiment voluntarily, under controlled conditions and in collaboration with the Food and Drug Administration, with graphics and other formats for presenting nutrition and other related food labeling information that is consistent with the current quantitative system in *§§101.9* and *105.66* of this chapter.

b) Any firm that intends to undertake a labeling experiment that requires exemptions from certain requirements of *§§101.9* and *105.66* of this chapter should submit a written proposal containing a thorough discussion of each of the following information items that apply to the particular experiment:
   (1) A description of the labeling format to be tested;
   (2) A statement of the criteria to be used in the experiment for assigning foods to categories, e.g., nutrient or other values defining "*low*" and "*reduced*";
   (3) A draft of the material to be used in the store, e.g., shelf tags, booklets, posters, etc.;
   (4) The dates on which the experiment will begin and end and on which a written report of analysis of the experimental data will be submitted to FDA, together with a commitment not to continue the experiment beyond the proposed ending date without FDA approval;
   (5) The geographic area or areas in which the experiment is to be conducted;
   (6) The mechanism to measure the effectiveness of the experiment;
   (7) The method for conveying to consumers the required nutrition and other labeling information that is exempted from the label during the experiment;
   (8) The method that will be or has been used to determine the actual nutritional characteristics of foods for which a claim is made; and
   (9) A statement of the sections of the regulations for which an exemption is sought.

c) The written proposal should be sent to the Dockets Management Branch (HFA-305), Food and Drug Administration, rm.1-23, 12420 Parklawn Dr., Rockville, MD 20857. The proposal should be clearly identified as a request for a temporary exemption for purposes of conducting authorized food labeling experiments and submitted as a citizen petition under *§10.30* of this chapter.

d) Approval for food labeling experiments will be given by FDA in writing. Foods labeled in violation of existing regulations will be subject to regulatory action unless an FDA-approved exemption to the specific regulation has been granted for that specific product.

e) Reporting requirements contained in *§101.108(b)* have been approved by this Office of Management and Budget and assigned number 0910-0151.

APPENDIX A TO PART 101—Monier-Williams Procedure (With Modification) for Sulfites in Food, Center for Food Safety and Applied Nutrition, Food and Drug Administration (November 1985)

The AOAC official method for sulfites *(Official Methods of Analysis, 14th Edition, 20.123-20.125, Association of Official Analytical Chemists)* has been modified, in FDA laboratories, to facilitate the determination of sulfites at or near 10 ppm in food. Method instructions, including modifications, are described below.

Apparatus—The apparatus (see *21 CFR 101.108* p.163 (4/1/04)) is designed to accomplish the selective transfer of sulfur dioxide from the sample in boiling aqueous hydrochloric acid to a solution of 3% hydrogen peroxide. This apparatus is easier to assemble than the official apparatus and the back pressure inside the apparatus is limited to the unavoidable pressure due to the height of the 3% H_2O_2

Chapter VIII

solution above the tip of the bubbler (F). Keeping the back pressure as low as possible reduces the likelihood that sulfur dioxide will be lost through leaks.

The apparatus should be assembled with a thin film of stopcock grease on the sealing surfaces of all the joints except the joint between the separatory funnel and the flask. Each joint should be clamped together to ensure a complete seal throughout the analysis. The separatory funnel, B, should have a capacity of 100 ml or greater. An inlet adapter, A, with a hose connector (Kontes K-183000 or equivalent) is required to provide a means of applying a head of pressure above the solution. (A pressure equalizing dropping funnel is not recommended because condensate, perhaps with sulfur dioxide, is deposited in the funnel and the side arm.) The round bottom flask, C, is a 1,000 ml flask with three 24/40 tapered joints. The gas inlet tube, D, (Kontes K-179000 or equivalent) should be of sufficient length to permit introduction of the nitrogen within 2.5 cm of the bottom of the flask. The Allihn condenser, E, (Kontes K431000-2430 or equivalent) has a jacket length of 300 mm. The bubbler, F, was fabricated from glass (see *21 CFR 101.108* p.163 (4/1/04)). The 3% hydrogen peroxide solution can be contained in a vessel, G, with an i.d. of ca. 2.5 cm and a depth of 18 cm.

Buret—A 10 ml buret (Fisher Cat. No. 03-848-2A or equivalent) with overflow tube and hose connections for an Ascarite tube or equivalent air scrubbing apparatus. This will permit the maintenance of a carbon dioxide free atmosphere over the standardized 0.01N sodium hydroxide.

Chilled Water Circulator—The condenser must be chilled with a coolant, such as 20% methanol-water maintained at 5 ° C. A circulating pump equivalent to the Neslab Coolflow 33 is suitable.

Reagents

a) *Aqueous hydrochloric acid, 4N.*—For each analysis prepare 90 ml of hydrochloric acid by adding 30 ml of concentrated hydrochloric acid (12N) to 60 ml of distilled water.
b) *Methyl red indicator*—Dissolve 250 mg of methyl red in 100 ml ethanol.
c) *Hydrogen peroxide solution, 3%*—Dilute ACS reagent grade 30% hydrogen peroxide to 3% with distilled water. Just prior to use, add three drops of methyl red indicator and titrate to a yellow end-point using 0.01N sodium hydroxide. If the end-point is exceeded discard the solution and prepare another 3% H_2O_2 solution.
d) *Standardized titrant, 0.01N NaOH*—Certified reagent may be used (Fisher S0-5-284). It should be standardized with reference standard potassium hydrogen phthalate.
e) *Nitrogen*—A source of high purity nitrogen is required with a flow regulator that will maintain a flow of 200 cc per minute. To guard against the presence of oxygen in the nitrogen, an oxygen scrubbing solution such as an alkaline pyrogallol trap may be used. Prepare pyrogallol trap as follows:
   1. Add 4.5 g pyrogallol to the trap.
   2. Purge trap with nitrogen for 2 to 3 minutes.
   3. Prepare a KOH solution prepared by adding 65g KOH to 85 ml distilled water (caution: heat).
   4. Add the KOH solution to the trap while maintaining an atmosphere of nitrogen in the trap.

Determination

Assemble the apparatus (see *21 CFR 101.108* p.163 (4/1/04)). The flask C must be positioned in a heating mantle that is controlled by a power regulating device such as Variac or equivalent. Add 400 ml of distilled water to flask C. Close the stopcock of separatory funnel, B, and add 10 ml of 4N hydrochloric acid to the separatory funnel. Begin the flow of nitrogen at a rate of 200±10 cc/min. The condenser coolant flow must be initiated at this time. Add 30 ml of 3% hydrogen peroxide, which has been titrated to a yellow end-point with 0.01N NaOH, to container G. After fifteen minutes the apparatus and the distilled water will be thoroughly de-oxygenated and the apparatus is ready for sample introduction.

Sample preparation (solids)—Transfer 50 g of food, or a quantity of food with a convenient quantity of SO_2 (500 to 1,500 mcg SO_2), to a food processor or blender. Add 100 ml of 5% ethanol in water and briefly grind the mixture. Grinding or blending should be continued only until the food is chopped into pieces small enough to pass through the 24/40 point of flask C.

Sample preparation (liquids)—Mix 50 g of the sample, or a quantity with a convenient quantity of SO_2 (500 to 1,500 mcg SO_2), with 100 ml of 5% ethanol in water.

Sample introduction and distillation—Remove the separatory funnel, B. and quantitatively transfer the food sample in aqueous ethanol to flask C. Wipe the tapered joint clean with a laboratory tissue, apply stopcock grease to the outer joint of the separatory funnel, and return the separatory funnel, B, to tapered joint flask C. The nitrogen flow through the 3% hydrogen peroxide solution should resume as soon as the funnel, B, is re-inserted into the appropriate joint in flask C. Examine each joint to ensure that it is sealed.

Apply a head pressure above the hydrochloric acid solution in B with a rubber bulb equipped with a valve. Open the stopcock in B and permit the hydrochloric acid solution to flow into flask C. Continue to maintain sufficient pressure above the acid solution to force the solution into the flask C. The stopcock may be closed, if necessary, to pump up the pressure above the acid and then opened again. Close the stopcock before the last few milliliters drain out of the separatory funnel, B, to guard against the escape of sulfur dioxide into the separatory funnel.

Apply the power to the heating mantle. Use a power setting which will cause 80 to 90 drops per minute of condensate to return to the flask from condenser, E. After 1.75 hours of boiling the contents of the 1,000 ml flask remove trap G.

Titration.—Titrate the contents with $0.01N$ sodium hydroxide. Titrate with $0.01N$ NaOH to a yellow endpoint that persists for at least twenty seconds. Compute the sulfite content, expressed as micrograms sulfur dioxide per gram of food (ppm) as follows:

$$ppm = (32.03 \times V_B \times N \times 1,000) + Wt$$

where 32.03 = milliequivalent weight of sulfur dioxide; V_B = volume of sodium hydroxide titrant of normality, N, required to reach endpoint; the factor, 1,000, converts milliequivalents to microequivalents and Wt = weight (g) of food sample introduced into the 1,000 ml flask.

Section F: Temporary Permits for Interstate Shipment of Experimental Packs of Food Varying from the Requirements of Definitions and Standards of Identity *(21 CFR 130.17)*

a) The Food and Drug Administration recognizes that before petitions to amend food standards can be submitted, appropriate investigations of potential advances in food technology sometimes require tests in interstate markets of the advantages to and acceptance by consumers of experimental packs of food varying from applicable definitions and standards of identity prescribed under section 401 of the act.

b) It is the purpose of the Food and Drug Administration to permit such tests when it can be ascertained that the sole purpose of the tests is to obtain data necessary for reasonable grounds in support of a petition to amend food standards, that the tests are necessary to the completion or conclusiveness of an otherwise adequate investigation, and that the interests of consumers are adequately safeguarded; permits for such tests shall normally be for a period not to exceed 15 months. The Food and Drug Administration, or good cause shown by the applicant, may provide for a longer test market period. The Food and Drug Administration will therefore refrain from recommending regulatory proceedings under the act on the charge that a food does not conform to an applicable standard, if the person who introduces or causes the introduction of the food into interstate commerce holds an effective permit from the Food and Drug Administration providing specifically for those variations in respect to which the food fails to conform to the applicable definition and standard of identity. The test period will begin on the date the person holding an effective permit from the Food and Drug Administration introduces or causes the introduction of the food covered by the permit into interstate commerce but not later than 3 months after notice of the issuance of the permit is published in the *FEDERAL REGISTER*. The Food and Drug Administration shall be notified in writing of the date on which the test period begins as soon as it is determined.

c) Any person desiring a permit may file with the Team Leader for Conventional Foods, Office of Nutritional Products, Labeling, and Dietary Supplements, Center for Food Safety and Applied Nutrition (HFS-822), 200 C St. SW., Washington, DC 20204, a written application in triplicate containing as part thereof the following:

(1) Name and address of the applicant.

(2) A statement of whether or not the applicant is regularly engaged in producing the food involved.

(3) A reference to the applicable definition and standard of identity (citing applicable section of regulations).

(4) A full description of the proposed variation from the standard.

(5) The basis upon which the food so varying is believed to be wholesome and nondeleterious.

(6) The amount of any new ingredient to be added; the amount of any ingredient, required by the standard, to be eliminated; any change of concentration not contemplated by the standard; or any change in name that would more appropriately describe the new product under test. If such new ingredient is not a commonly known food ingredient, a description of its properties and basis for concluding that it is not a deleterious substance.

(7) The purpose of effecting the variation.

(8) A statement of how the variation is of potential advantage to consumers. The statement shall include the reasons why the applicant does not consider the data obtained in any prior investigations which may have been conducted sufficient to support a petition to amend the standard.

(9) The proposed label (or an accurate draft) to be used on the food to be market tested. The label shall conform in all respects to the general requirements of the act and shall provide a means whereby the consumer can distinguish between the food being tested and such food complying with the standard.

(10) The period during which the applicant desires to introduce such food into interstate commerce, with a statement of the reasons supporting the need for such period. If a period longer than 15 months is requested, a detailed explanation of why a 15-month period is inadequate shall be provided.

(11) The probable amount of such food that will be distributed. The amount distributed should be limited to the smallest number of units reasonably required for a bona fide market test. Justification for the amount requested shall be included.

(12) The areas of distribution.

(13) The address at which such food will be manufactured.

(14) A statement of whether or not such food has been or is to be distributed in the State in which it was manufactured.

(15) If it has not been or is not to be so distributed, a statement showing why.

(16) If it has been or is to be so distributed, a statement of why it is deemed necessary to distribute such food in other States.

d) The Food and Drug Administration may require the applicant to furnish samples of the food varying from the standard and to furnish such additional information as may be deemed necessary for action on the application.

e) If the Food and Drug Administration concludes that the variation may be advantageous to consumers and will not result in failure of the food to conform to any provision of the act except section 403(g), a permit shall be issued to the applicant for interstate shipment of such food. The terms and conditions of the permit shall be those set forth in the application with such modifications, restrictions, or qualifications as the Food and Drug Administration may deem necessary and state in the permit.

f) The terms and conditions of the permit may be modified at the discretion of the Food and Drug Administration or upon application of the permittee during the effective period of the permit.

g) The Food and Drug Administration may revoke a permit for cause which shall include but not be limited to the following:

(1) That the permittee has introduced a food into interstate commerce contrary to the terms and conditions of the permit.

(2) That the application for a permit contains an untrue statement of a material fact.

(3) That the need therefor no longer exists.

h) During the period within which any permit is effective, it shall be deemed to be included within the terms of any guaranty or undertaking otherwise effective pursuant to the provisions of section 303(c) of the act.

i) If an application is made for an extension of the permit, it shall be accompanied by a description of experiments conducted under the permit, tentative conclusions reached, and reasons why further experimental shipments are considered necessary. The application for an extension shall be filed not later than 3 months prior to the expiration date of the permit and shall be accompanied by a petition to amend the affected food standard. If the Food and Drug Administration concludes that it will be in the interest of consumers to issue an extension of the time period for the market test, a notice will be published in the *FEDERAL REGISTER* stating that fact. The notice will include an invitation to all interested persons to participate in the market test under the same conditions that applied to the initial permit holder, including labeling and the amount to be distributed, except that the designated area of distribution shall not apply. The extended market test period shall not begin prior to the publication of a notice in the *FEDERAL REGISTER* granting the extension and shall terminate either on the effective date of an affirmative order ruling on the proposal or 30 days after a negative order ruling on the proposal, whichever the case may be. Any interested person who accepts the invitation to participate in the extended market test shall notify the Food and Drug Administration in writing of that fact, the amount to be distributed, and the area of distribution; and along with such notification, he shall submit the labeling under which the food is to be distributed.

j) Notice of the granting or revocation of any permit shall be published in the *FEDERAL REGISTER*.

k) All applications for a temporary permit, applications for an extension of a temporary permit, and related records are available for public disclosure when the notice of a permit or extension thereof is published in the *FEDERAL REGISTER*. Such disclosure shall be in accordance with the rules established in part 20 of this chapter.

l) Any person who contests denial, modification, or revocation of a temporary permit shall have an opportunity for a regulatory hearing before the Food and Drug Administration pursuant to part 16 of this chapter.

Chapter IX
Compliance Provisions

Section A: Failure to Reveal Material Facts *(21 CFR 1.21)*

(a) Labeling of a food, drug, device or cosmetic shall be deemed to be misleading if it fails to reveal facts that are:
    (1) Material in light of other representations made or suggested by statement, word, design, device or any combination thereof; or
    (2) Material with respect to consequences which may result from use of the article under:
        (i) The conditions prescribed in such labeling; or
        (ii) such conditions of use as are customary or usual.
(b) Affirmative disclosure of material facts pursuant to paragraph (a) of this section may be required, among other appropriate regulatory procedures, by
    (1) Regulations in this chapter promulgated pursuant to section 701(a) of the act; or
    (2) Direct court enforcement action.
(c) Paragraph (a) of this section does not:
    (1) Permit a statement of differences of opinion with respect to warnings (including contraindications, precautions, adverse reactions, and other information relating to possible product hazards) required in labeling for food, drugs, devices, or cosmetics under the act.
    (2) Permit a statement of differences of opinion with respect to the effectiveness of a drug unless each of the opinions expressed is supported by substantial evidence of effectiveness as defined in sections 505(d) and 512(d) of the act.

Section B: Misleading Containers *(21 CFR 100.100)*

In accordance with section 403(d) of the act, a food shall be deemed to be misbranded if its container is so made, formed, or filled as to be misleading.

(a) A container that does not allow the consumer to fully view its contents shall be considered to be filled as to be misleading if it contains nonfunctional slack-fill. Slack-fill is the difference between the actual capacity of a container and the volume of product contained therein. Nonfunctional slack-fill is the empty space in a package that is filled to less than its capacity for reasons other than:
    (1) Protection of the contents of the package;
    (2) The requirements of the machines used for enclosing the contents in such package;
    (3) Unavoidable product settling during shipping and handling;
    (4) The need for the package to perform a specific function (e.g., where packaging plays a role in the preparation or consumption of a food), where such function is inherent to the nature of the food and is clearly communicated to consumers;

(5) The fact that the product consists of a food packaged in a reusable container where the container is part of the presentation of the food and has value which is both significant in proportion to the value of the product and independent of its function to hold the food, e.g., a gift product consisting of a food or foods combined with a container that is intended for further use after the food is consumed; or durable commemorative or promotional packages; or

(6) Inability to increase level of fill or to further reduce the size of the package (e.g., where some minimum package size is necessary to accommodate required food labeling (excluding any vignettes or other nonmandatory designs or label information), discourage pilfering, facilitate handling, or accommodate tamper-resistant devices).

(b) [Reserved]

Section C: Food with a Label Declaration of Nutrients *(21 CFR 101.9(g))*

(g) Compliance with this section shall be determined as follows:

(1) A collection of primary containers or units of the same size, type, and style produced under conditions as nearly uniform as possible, designated by a common container code or marking, or in the absence of any common container code or marking, a day's production, constitutes a "lot."

(2) The sample for nutrient analysis shall consist of a composite of 12 sub-samples (consumer units), taken 1 from each of 12 different randomly chosen shipping cases, to be representative of a lot. Unless a particular method of analysis is specified in paragraph (c) of this section, composites shall be analyzed by appropriate methods as given in the "*Official Methods of Analysis of the AOAC International,*" 15th Ed. (1990), which is incorporated by reference in accordance with 5 U.S.C. 552(a) or *1 CFR part 51* or, if no AOAC method is available or appropriate, by other reliable and appropriate analytical procedures.

(3) Two classes of nutrients are defined for purposes of compliance:
    (i) *Class I.* Added nutrients in fortified or fabricated foods; and
    (ii) *Class II.* Naturally occurring (indigenous) nutrients. If any ingredient which contains a naturally occurring (indigenous) nutrient is added to a food, the total amount of such nutrient in the final food product is subject to class II requirements unless the same nutrient is also added.

(4) A food with a label declaration of a vitamin, mineral, protein, total carbohydrate dietary fiber, other carbohydrate, polyunsaturated or monounsaturated fat, or potassium shall be deemed to be misbranded under section 403(a) of the Federal Food, Drug, and Cosmetic Act (the act) unless it meets the following requirements:
    (i) *Class I vitamin, mineral, protein, dietary fiber, or potassium.* The nutrient content of the composite is at least equal to the value for that nutrient declared on the label.
    (ii) *Class II vitamin, mineral, protein, total carbohydrate, dietary fiber, other carbohydrate, polyunsaturated or monounsaturated fat, or potassium.* The nutrient content of the composite is at least equal to 80 percent of the value for that nutrient declared on the label.
    Provided, That no regulatory action will be based on a determination of a nutrient value that falls below this level by a factor less than the variability generally recognized for the analytical method used in that food at the level involved.

(5) A food with a label declaration of calories, sugars, total fat, saturated fat, cholesterol, or sodium shall be deemed to be misbranded under section 403(a) of the act if the nutrient content of the composite is greater than 20 percent in excess of the value for that nutrient declared on the label. *Provided,* That no regulatory action will be based on a determination of a nutrient value that falls above this level by a factor less than the variability generally recognized for the analytical method used in that food at the level involved.

(6) Reasonable excesses of a vitamin, mineral, protein, total carbohydrate, dietary fiber, other carbohydrate, polyunsaturated or monounsaturated fat, or potassium over labeled amounts are acceptable within current good manufacturing practice. Reasonable deficiencies of calories, sugars, total fat, saturated fat, cholesterol, or sodium under labeled amounts are acceptable within current good manufacturing practice.

(7) Compliance will be based on the metric measure specified in the label statement of serving size.

(8) Compliance with the provisions set forth in paragraphs (g)(1) through (g)(6) of this section may be provided by use of an FDA approved data base that has been computed following FDA guideline procedures and where food samples have been handled in accordance with current good manufacturing practice to prevent nutrition loss. FDA approval of a database shall not be considered granted until the Center for Food Safety and Applied Nutrition has agreed to all aspects of the database in writing. The approval will be granted where a clear need is presented (e.g., raw produce and seafood). Approvals will be in effect for a limited time, e.g., 10 years, and will be eligible for renewal in the absence of significant changes in agricultural or industry practices. Approval requests shall be submitted in accordance with the provisions of §10.30 of this chapter. Guidance in the use of data bases may be found in the *"FDA Nutrition Labeling Manual--A Guide for Developing and Using Data Bases,"* available from the Office of Nutritional Products, Labeling and Dietary Supplements, (HFS-800), Center for Food Safety and Applied Nutrition, Food and Drug Administration, 5100 Paint Branch Pkwy., College Park, MD 20740.

(9) When it is not technologically feasible, or some other circumstance makes it impracticable, for firms to comply with the requirements of this section (e.g., to develop adequate nutrient profiles to comply with the requirements of paragraph (c) of this section), FDA may permit alternative means of compliance or additional exemptions to deal with the situation. Firms in need of such special allowances shall make their request in writing to the Office of Nutritional Products, Labeling and Dietary Supplements, (HFS-800), Center for Food Safety and Applied Nutrition, Food and Drug Administration, 5100 Paint Branch Pkwy., College Park, MD 20740.

Section D: Food Subject to Nutrition Labeling *(21 CFR 101.9(k))*

(k) A food labeled under the provisions of this section shall be deemed to be misbranded under sections 201(n) and 403(a) of the act if its label or labeling represents, suggests, or implies:

(1) That the food, because of the presence or absence of certain dietary properties, is adequate or effective in the prevention, cure, mitigation, or treatment of any disease or symptom. Information about the relationship of a dietary property to a disease or health related condition may only be provided in conformance with the requirements of §101.14 and *part 101, subpart E.*

(2) That the lack of optimum nutritive quality of a food, by reason of the soil on which that food was grown, is or may be responsible for an inadequacy or deficiency in the quality of the daily diet.

(3) That the storage, transportation, processing, or cooking of a food is or may be responsible for an inadequacy or deficiency in the quality of the daily diet.

(4) That a natural vitamin in a food is superior to an added or synthetic vitamin or to differentiate in any way between vitamins naturally present from those added.

Section E: Food: Prominence of Required Statements *(21 CFR 101.15)*

(a) A word, statement, or other information required by or under authority of the act to appear on the label may lack that prominence and conspicuousness required by section 403(f) of the act by reason (among other reasons) of:

(1) The failure of such word, statement, or information to appear on the part or panel of the label which is presented or displayed under customary conditions of purchase;

(2) The failure of such word, statement, or information to appear on two or more parts or panels of the label, each of which has sufficient space therefor, and each of which is so designed as to render it likely to be, under customary conditions of purchase, the part or panel displayed;

(3) The failure of the label to extend over the area of the container or package available for such extension, so as to provide sufficient label space for the prominent placing of such word, statement, or information;

(4) Insufficiency of label space (for the prominent placing of such word, statement, or information) resulting from the use of label space for any word, statement, design, or device which is not required by or under authority of the act to appear on the label;

(5) Insufficiency of label space (for the prominent placing of such word, statement, or information) resulting from the use of label space to give materially greater conspicuousness to any other word, statement, or information, or to any design or device; or

(6) Smallness or style of type in which such word, statement, or information appears, insufficient

background contrast, obscuring designs or vignettes, or crowding with other written, printed, or graphic matter.

(b) No exemption depending on insufficiency of label space, as prescribed in regulations promulgated under section 403(e) or (i) of the act, shall apply if such insufficiency is caused by:

    (1) The use of label space for any word, statement, design, or device which is not required by or under authority of the act to appear on the label;

    (2) The use of label space to give greater conspicuousness to any word, statement, or other information than is required by section 403(f) of the act; or

    (3) The use of label space for any representation in a foreign language.

(c) (1) All words, statements, and other information required by or under authority of the act to appear on the label or labeling shall appear thereon in the English language: *Provided, however,* That in the case of articles distributed solely in the Commonwealth of Puerto Rico or in a Territory where the predominant language is one other than English, the predominant language may be substituted for English.

    (2) If the label contains any representation in a foreign language, all words, statements, and other information required by or under authority of the act to appear on the label shall appear thereon in the foreign language: *Provided, however,* That individual serving size packages of foods containing no more than 1-1/2 avoirdupois ounces or no more than 1-1/2 fluid ounces served with meals in restaurants, institutions, and passenger carriers and not intended for sale at retail are exempt from the requirements of this paragraph (c)(2), if the only representation in the foreign language(s) is the name of the food.

    (3) If any article of labeling (other than a label) contains any representation in a foreign language, all words, statements, and other information required by or under authority of the act to appear on the label or labeling shall appear on such article of labeling.

Section F: Misbranding of Food *(21 CFR 101.18)*

(a) Among representations in the labeling of a food which render such food misbranded is a false or misleading representation with respect to another food or a drug, device, or cosmetic.

(b) The labeling of a food which contains two or more ingredients may be misleading by reason (among other reasons) of the designation of such food in such labeling by a name which includes or suggests the name of one or more but not all such ingredients, even though the names of all such ingredients are stated elsewhere in the labeling.

(c) Among representations in the labeling of a food which render such food misbranded is any representation that expresses or implies a geographical origin of the food or any ingredient of the food except when such representation is either:

    (1) A truthful representation of geographical origin.

    (2) A trademark or trade name provided that as applied to the article in question its use is not deceptively misdescriptive. A trademark or trade name composed in whole or in part of geographical words shall not be considered deceptively misdescriptive if it:

        (i) Has been so long and exclusively used by a manufacturer or distributor that it is generally understood by the consumer to mean the product of a particular manufacturer or distributor; or

        (ii) Is so arbitrary or fanciful that it is not generally understood by the consumer to suggest geographic origin.

    (3) A part of the name required by applicable Federal law or regulation.

    (4) A name whose market significance is generally understood by the consumer to connote a particular class, kind, type, or style of food rather than to indicate geographical origin.

Chapter IX

Section G: Substantial Compliance of Food Retailers with the Guidelines for the Voluntary Nutrition Labeling of Raw Fruits, Vegetables, and Fish *(21 CFR 101.43)*

(a) The Food and Drug Administration (FDA) will judge a food retailer who sells raw agricultural commodities or raw fish to be in compliance with the guidelines in § 101.45 with respect to raw agricultural commodities if the retailer displays or provides nutrition labeling for at least 90 percent of the raw agricultural commodities listed in §101.44 that it sells, and with respect to raw fish if the retailer displays or provides nutrition labeling for at least 90 percent of the types of raw fish listed in §101.44 that it sells. To be in compliance, the nutrition labeling shall:

   (1) Be presented in the store or other type of establishment in a manner that is consistent with § 101.45(a)(1);

   (2) Be presented in content and format that are consistent with §101.45(a)(2), (a)(3), and (a)(4); and

   (3) Include data that have been provided by FDA in appendices C and D to part 101 of this chapter, except that potassium is voluntary.

(b) To determine whether there is substantial compliance by food retailers with the guidelines in § 101.45 for the voluntary nutrition labeling of raw fruit and vegetables and of raw fish, FDA will select a representative sample of 2,000 stores, allocated by store type and size, for raw fruit and vegetables and for raw fish.

(c) FDA will find that there is substantial compliance with the guidelines in § l01.45 if it finds based on paragraph (a) of this section that at least 60 percent of all stores that are evaluated are in compliance.

(d) FDA will evaluate substantial compliance separately for raw agricultural commodities and for raw fish.

Chapter X
Special Food Labeling Issues

Section A. FDA Letter on Labeling Food Products Presented or Available on the Internet

Daniel J. Popeo
Paul D. Kamenar
Washington Legal Foundation
2009 Massachusetts Avenue, N.W.
Washington, DC 20036

Re: Docket No. 01P-0187/CP 1

Dear Messers. Popeo and Kamenar:

This letter responds to your citizen petition, received by the Food and Drug Administration (FDA) on April 16, 2001, filed on behalf of the Washington Legal Foundation. Your petition asked FDA to "formally adopt a rule, policy, or guidance stating that information presented or available on a company's Internet Web site, including hyperlinks to other third party sites, does not constitute 'labeling,'" as defined by the Federal Food, Drug, and Cosmetic Act (FDCA) at *21 U.S.C. § 321(m)*. In your petition, you further requested that the rule, policy, or guidance specify that such information may, but does not necessarily, constitute advertising. Alternatively, you asked FDA to adopt a rule, policy, or guidance "exempting Internet information of food companies from labeling requirements."

FDA agrees that Internet information, particularly those Web sites that provide truthful and non-misleading information about FDA-regulated products, can serve a valuable and useful function. The agency also agrees that it has not issued a specific rule, policy, or guidance that addresses whether information posted on a company's Web site is considered advertising, labeling, neither, or both. However, FDA disagrees that information presented or available on a company's website could never constitute labeling. "Labeling" is defined in section 201(m) of the FDCA (21 U.S.C. § 321(m)) as "all labels and other written, printed or graphic matter upon any article... or accompanying such article." In *Kordel v. United States*, 335 U.S. 345 (1948), the Supreme Court concluded that the phrase "accompanying such article" included literature that was shipped separately and at different times from the drugs with which they were associated. "One article or thing is accompanied by another when it supplements or explains it, in the manner that a committee report of the Congress accompanies a bill. No physical attachment one to the other is necessary. It is the textual relationship that is significant." Id. at 350. The Court also noted that the literature and drugs were parts of an integrated distribution program.

Based on this authority, FDA and the courts have interpreted "labeling" to include "[b]rochures, booklets, ... motion picture films, film strips, ... sound recordings, ... and similar pieces of printed, audio, or visual matter descriptive of a drug... which are disseminated by or on behalf of its manufacturer, packer, or distributor...." *21 C.F.R. § 202.1(l)(2)*; See *SmithKline Beecham Consumer Healthcare, L.P. v. Watson Pharms., Inc.*, 211 F. 3d 21, 26 (2d Cir. 2000) (dictum) (copyrighted user's guide and audiotape for nicotine gum constitute "labeling").

Lower court cases after *Kordel* reinforce a broad reading of the term "accompanying." See *United States v. Diapulse Manufacturing Corp. of America*, 389 F. 2d 612 (2d Cir. 1968); *V.E. Irons, Inc. v. United States*, 244 F. 2d 34 (1st Cir. 1957), cert. denied, 77 S. Ct. 1383 (1957). In addition, the courts have considered whether the information and the product are part of an integrated distribution program, where, for example, the information and the product originate from the same source or the information is designed to promote the distribution and sale of the product, even if such sale is not immediate. See *United States v. 47 Bottles, More or Less, Jenasol RJ Formula "60"*, 320 F. 2d 564 (3d Cir. 1963); *United States v. Guardian Chemical*, 410 F.2d 157 (2d Cir. 1969).

Accordingly, FDA believes that, in certain circumstances, information about FDA-regulated products that is disseminated over the Internet by, or on behalf of, a regulated company can meet the definition of labeling in section 201(m) of the FDCA. For example, if a company were to promote a regulated product on its Web site and allow consumers to purchase the product directly from the web site, the web site is likely to be "labeling." The Web site, in that case, would be written, printed, or graphic matter that supplements or explains the product and is designed for use in the distribution and sale of the product.

To provide an example from the other end of the spectrum, some product-specific promotion presented on non-company websites that is very much similar, if not identical, to messages the agency has traditionally regulated as advertisements in print media (e.g., advertisements published in journals, magazines, periodicals, and newspapers) would be viewed as advertising. These are just examples at the extremes and, as discussed below, the agency will proceed on a case-by-case basis in determining what is "labeling."

The agency sees no reason to treat Internet information of food companies differently from Internet information of other FDA-regulated industries. As such, FDA disagrees with your alternative request to exempt Internet information of food companies from labeling requirements.

Government agencies possess broad discretion in deciding whether to proceed by general rulemaking or case-by-case adjudication. *NLRB v. Bell Aerospace Co., Div. of Textron, Inc.* 416 U.S. 267, 294 (1974); *SEC v. Chenery Corp.*, 332 U.S. 194, 203 (1947); *Teva Pharmaceuticals, USA, Inc. v. FDA*, 182 F. 3d 1003, 1010 (D.C. Cir. 1999). FDA has explored developing a guidance on promotion of FDA-regulated products on the Internet, but has decided not to issue a document at this time. The agency believes that any rule or guidance on this issue would be quickly outdated due to the ongoing rapid changes in the Internet and its use. As a result, issuing a rule or guidance may stifle innovation and create greater confusion among industry and the public. Therefore, for the time being, FDA will continue to use a case-by-case approach based on the specific facts of each case.

Although for the reasons stated above, FDA has decided to deny your petition, generally, at a company's request, the agency is willing to discuss a company's specific plans for posting information on its Web site or linking to information on a third-party Web site.
FDA appreciates your interest in this area.

Sincerely yours,

Margaret M. Dotzel
Associate Commissioner for Policy

Adopted from FDA, CFSAN, Overview of Dietary Supplements, 1/3/01

Chapter X

Section B. FDA/FTC Labeling Responsibility

Under DSHEA, a firm is responsible for determining that the dietary supplements it manufactures or distributes are safe and that any representations or claims made about them are substantiated by adequate evidence to show that they are not false or misleading. This means that dietary supplements do not need approval from FDA before they are marketed. Except in the case of a new dietary ingredient, where pre-market review for safety data and other information is required by law, a firm does not have to provide FDA with the evidence it relies on to substantiate safety or effectiveness before or after it markets its products.

Also, manufacturers do not need to register themselves nor their dietary supplement products with FDA before producing or selling them. Currently, there are no FDA regulations that are specific to dietary supplements that establish a minimum standard of practice for manufacturing dietary supplements. However, FDA intends to issue regulations on good manufacturing practices that will focus on practices that ensure the identity, purity, quality, strength and composition of dietary supplements. At present, the manufacturer is responsible for establishing its own manufacturing practice guidelines to ensure that the dietary supplements it produces are safe and contain the ingredients listed on the label.

By law (DSHEA), the manufacturer is responsible for ensuring that its dietary supplement products are safe before they are marketed. Unlike drug products that must be proven safe and effective for their intended use before marketing, there are no provisions in the law for FDA to "approve" dietary supplements for safety or effectiveness before they reach the consumer. Also unlike drug products, manufacturers and distributors of dietary supplements are not currently required by law to record, investigate or forward to FDA any reports they receive of injuries or illnesses that may be related to the use of their products. Under DSHEA, once the product is marketed, FDA has the responsibility for showing that a dietary supplement is "unsafe," before it can take action to restrict the product's use or removal from the marketplace.

Because dietary supplements are under the "umbrella" of foods, FDA's Center for Food Safety and Applied Nutrition (CFSAN) is responsible for the agency's oversight of these products. FDA's efforts to monitor the marketplace for potential *illegal* products (that is, products that may be unsafe or make false or misleading claims) include obtaining information from inspections of dietary supplement manufacturers and distributors, the Internet, consumer and trade complaints, occasional laboratory analyses of selected products, and adverse events associated with the use of supplements that are reported to the agency.

FDA receives many consumer inquiries about the validity of claims for dietary supplements, including product labels, advertisements, media, and printed materials. The responsibility for ensuring the validity of these claims rests with the manufacturer, FDA, and, in the case of advertising, with the Federal Trade Commission.

By law, manufacturers may make three types of claims for their dietary supplement products: health claims, structure/function claims, and nutrient content claims. Some of these claims describe: the link between a food substance and disease or a health-related condition; the intended benefits of using the product; or the amount of a nutrient or dietary substance in a product. Different requirements generally apply to each type of claim.

The Federal Trade Commission (FTC) regulates advertising, including infomercials, for dietary supplements and most other products sold to consumers. FDA works closely with FTC in this area, but FTC's work is directed by different laws. Advertising and promotional material received in the mail are also regulated under different laws and are subject to regulation by the U.S. Postal Inspection Service.

The role of the Federal Trade Commission, which enforces laws outlawing "unfair or deceptive acts or practices," is to ensure that consumers get accurate information about dietary supplements so that they can make informed decisions about these products.

The Federal Trade Commission (FTC) and the Food and Drug Administration (FDA) work together under a long-standing liaison agreement governing the division of responsibilities between the two agencies. As applied to dietary supplements, the FDA has primary responsibility for claims on product <u>labeling</u>, including packaging, inserts, and other promotional materials distributed at the point of sale. The FTC has primary responsibility for claims in <u>advertising</u>, including print and broadcast ads, infomercials, catalogs, and similar direct marketing materials. Marketing on the Internet is subject to regulation in the same fashion as promotions through any other media. Because of their shared jurisdiction, the two agencies work closely to ensure that their enforcement efforts are consistent to the fullest extent feasible.

In 1994, the Dietary Supplements Health and Education Act (DSHEA) significantly changed the FDA's role in regulating supplement labeling. Although DSHEA does not directly apply to advertising, it has generated many questions about the FTC's approach to dietary supplement advertising. The answer to these questions is that advertising for <u>any</u> product – including dietary supplements – must be truthful, not misleading, and substantiated. Given the dramatic increase in the volume and variety of dietary supplement advertising in recent years, FTC staff is issuing this guide to clarify how long-standing FTC policies and enforcement practices relate to dietary supplement advertising.

The FTC's approach to supplement advertising is best illustrated by its Enforcement Policy Statement on Food Advertising (Food Policy Statement). Although the Food Policy Statement does not specifically refer to supplements, the principles underlying the FTC's regulation of health claims in food advertising are relevant to the agency's approach to health claims in supplement advertising. In general, the FTC gives great deference to an FDA determination of whether there is adequate support for a health claim. Furthermore, the FTC and the FDA will generally arrive at the same conclusion when evaluating unqualified health claims. As the Food Policy Statement notes, however, there may be certain limited instances when a carefully qualified health claim in advertising may be permissible under FTC law, in circumstances where it has not been authorized for labeling. However, supplement marketers are cautioned that the FTC will require both strong scientific support and careful presentation for such claims. Supplement marketers should ensure that anyone involved in promoting products is familiar with basic FTC advertising principles. The FTC has taken action not just against supplement manufacturers, but also, in appropriate circumstances, against ad agencies, distributors, retailers, catalog companies, infomercial producers and others involved in deceptive promotions. *Therefore, all parties who participate directly or indirectly in the marketing of dietary supplements have an obligation to make sure that claims are presented truthfully and to check the adequacy of the support behind those claims.*

Adopted from FDA, CFSAN, Overview of Dietary Supplements, 1/3/01; and the FTC, BCP, Dietary Supplements: An Advertising Guide for Industry, 4/01, updated 7/16/03

Section C. Dietary Supplement /Drug Combination Product

1. Letter about the Regulatory Status of the _____ Pill

Dear _____:

This is to inform you of the regulatory status of the _____ Pill, a product that combines aspirin, an over-the-counter drug, with vitamin B-12, a dietary supplement. The purpose of this letter is to advise you that FDA has serious concerns about the marketing of any such combination product.

These combination products raise a number of significant public health and policy issues. For example, the addition of a new ingredient to a legally marketed drug product could affect the safety and efficacy of the drug component. In addition, consumers may be confused about the degree of scrutiny FDA gives such combination products. Consumers may believe that both components have been subjected to the more stringent drug regulatory requirements when, in fact, only the drug component may have been reviewed by the agency for safety and effectiveness. Moreover, it is uncertain under what circumstances the disclaimer required by the Dietary Supplement Health and Education Act (DSHEA) (codified in 21 U.S.C. 403(r)(6)(C)) could appear on a combination product without furthering consumer confusion.

The agency must determine under what conditions these combination products can be marketed in accordance with the Federal Food, Drug, and Cosmetic Act (FD&C Act), as amended by DSHEA. More specifically, the agency must determine what regulatory standards are appropriate, including, but not limited to, what safety and effectiveness standards will apply and how such products will be labeled. The number of inquiries we have received on this subject has made resolution of these issues a priority at the agency. We will be providing additional information as we develop our policy in this area.

Until the agency has carefully considered these issues, however, FDA strongly recommends that firms refrain from marketing products that combine both drug and dietary supplement ingredients (except for products marketed under an approved new drug application). In this interim period, we intend to take appropriate measures including, if necessary, regulatory action with respect to any such product that violates the FD&C Act or the agency's implementing regulations.

If you have any questions on this matter or wish to discuss it further with us, please contact Sharon Lindan Mayl at (301) 827-3360.

Sincerely yours,

Margaret M. Dotzel
Associate Commissioner for Policy

FDA, CFSAN, Letter about the Regulatory Status of the Good News Andy Pill, May 30, 2000

Section D: Food Bioengineering

The FDA issued a proposed rule (FR., Vol. 66, 1/18/01) and a draft guidance document (January 2001) concerning food developed through biotechnology. The proposed rule, if finalized, would require food developers to notify FDA at least 120 days in advance of their intent to market a food or animal feed developed through biotechnology and to provide information to demonstrate that the product is as safe as its conventional counterpart.

In a separate but related action, FDA issued a draft guidance document which if finalized, would provide direction to manufacturers who wish to label their food products as being made with or without ingredients developed through biotechnology. The draft Guidance Document is provided for guidance purposes.

On March 14, 2001, the agency updated its Internet site regarding completed consultations between FDA and developers of foods derived through biotechnology (also known as bioengineered foods) to include additional information about each completed consultation which is now available.

Guidance for Industry
Voluntary Labeling Indicating Whether Foods Have or Have Not Been Developed Using Bioengineering

Draft Guidance

This guidance document is being distributed for comment purposes only.

Draft released for comment January 2001.

Comments and suggestions regarding this draft document should be submitted to Dockets Management Branch (HFA-305), Food and Drug Administration, 5630 Fishers Lane, rm. 1061, Rockville, MD 20852. All comments should be identified with Docket Number 00D-1598. For questions regarding this draft document contact Catalina Ferre-Hockensmith, (202) 205-4168.

Guidance

In determining whether a food is misbranded, FDA would review label statements about the use of bioengineering to develop a food or its ingredients under sections 403(a) and 201(n) of the act. Under section 403(a) of the act, a food is misbranded if statements on its label or in its labeling are false or misleading in any particular. Under section 201(n), both the presence and the absence of information are relevant to whether labeling is misleading. That is, labeling may be misleading if it fails to disclose facts that are material in light of representations made about a product or facts that are material with respect to the consequences that may result from use of the product. In determining whether a statement that a food is or is not genetically engineered is misleading under sections 201(n) and 403(a) of the act, the agency will take into account the entire label and labeling.

Statements about foods developed using bioengineering

FDA recognizes that some manufacturers may want to use informative statements on labels and in labeling of bioengineered foods or foods that contain ingredients produced from bioengineered foods. The following are examples of some statements that might be used. The discussion accompanying each example is intended to provide guidance as to how similar statements can be made without being misleading.

- "Genetically engineered" or "This product contains cornmeal that was produced using biotechnology."

The information that the food was bioengineered is optional and this kind of simple statement is not likely to be misleading. However, focus group data indicate that consumers would prefer label statements that disclose and explain the goal of the technology (why it was used or what it does for/to the food) (Ref. 1). Consumers also expressed some preference for the term "biotechnology" over such terms as "genetic modification" and "genetic engineering" (Ref. 1).

- "This product contains high oleic acid soybean oil from soybeans developed using biotechnology to decrease the amount of saturated fat."

This example includes both required and optional information. As discussed above in the background section, when a food differs from its traditional counterpart such that the common or usual name no longer adequately describes the new food, the name must be changed to describe the difference. Because this soybean oil contains more oleic acid than traditional soybean oil, the term "soybean oil" no longer adequately describes the nature of the food. Under section 403(i) of the act, a phrase like "high oleic acid" would be required to appear as part of the name of the food to describe its basic nature. The statement that the soybeans were developed using biotechnology is optional. So is the statement that the reason for the change in the soybeans was to reduce saturated fat.

- "These tomatoes were genetically engineered to improve texture."

In this example, the change in texture is a difference that may have to be described on the label. If the texture improvement makes a significant difference in the finished product, sections 201(n) and 403(a)(1) of the act would require disclosure of the difference for the consumer. However, the statement must not be misleading. The phrase "to improve texture" could be misleading if the texture difference is not noticeable to the consumer. For example, if a manufacturer wanted to describe a difference in a food that the consumer would not notice when purchasing or consuming the product, the manufacturer should phrase the statements so that the consumer can understand the significance of the difference. If the change in the tomatoes was intended to facilitate processing but did not make a noticeable difference in the processed consumer product, a phrase like "to improve texture for processing" rather than "to improve texture" should be used to ensure that the consumer is not misled. The statement that the tomatoes were genetically engineered is optional.

- "Some of our growers plant tomato seeds that were developed through biotechnology to increase crop yield."

The entire statement in this example is optional information. The fact that there was increased yield does not affect the characteristics of the food and is therefore not necessary on the label to adequately describe the food for the consumer. A phrase like "to increase yield" should only be included where there is substantiation that there is in fact the stated difference.

Chapter X

Where a benefit from a bioengineered ingredient in a multi-ingredient food is described, the statement should be worded so that it addresses the ingredient and not the food as a whole; for example, "This product contains high oleic acid soybean oil from soybeans produced through biotechnology to decrease the level of saturated fat." In addition, the amount of the bioengineered ingredient in the food may be relevant to whether the statement is misleading. This would apply especially where the bioengineered difference is a nutritional improvement. For example, it would likely be misleading to make a statement about a nutritionally improved ingredient on a food that contains only a small amount of the ingredient, such that the food's overall nutritional quality would not be significantly improved.

FDA reminds manufacturers that the optional terms that describe an ingredient of a multi-ingredient food as bioengineered should not be used in the ingredient list of the multi-ingredient food. Section 403(i)(2) of the act requires each ingredient to be declared in the ingredient statement by its common or usual name. Thus, any terms not part of the name of the ingredient are not permitted in the ingredient statement. In addition, *21 CFR 101.2(e)* requires that the ingredient list and certain other mandatory information appear in one place without other intervening material. FDA has long interpreted any optional description of ingredients in the ingredient statement to be intervening material that violates this regulation.

Statements about foods that are not bioengineered or that do not contain ingredients produced from bioengineered foods

Terms that are frequently mentioned in discussions about labeling foods with respect to bioengineering include "GMO free" and "GM free." "GMO" is an acronym for "genetically modified organism" and "GM" means "genetically modified." Consumer focus group data indicate that consumers do not understand the acronyms "GMO" and "GM" and prefer label statements with spelled out words that mean bioengineering (Ref. 1).

Terms like "not genetically modified" and "GMO free," that include the word "modified" are not technically accurate unless they are clearly in a context that refers to bioengineering technology. "Genetic modification" means the alteration of the genotype of a plant using any technique, new or traditional. "Modification" has a broad context that means the alteration in the composition of food that results from adding, deleting, or changing hereditary traits, irrespective of the method. Modifications may be minor, such as a single mutation that affects one gene, or major alterations of genetic material that affect many genes. Most, if not all, cultivated food crops have been genetically modified. Data indicate that consumers do not have a good understanding that essentially all food crops have been genetically modified and that bioengineering technology is only one of a number of technologies used to genetically modify crops. Thus, while it is accurate to say that a bioengineered food was "genetically modified," it likely would be inaccurate to state that a food that had not been produced using biotechnology was "not genetically modified" without clearly providing a context so that the consumer can understand that the statement applies to bioengineering.

The term "GMO free" may be misleading on most foods, because most foods do not contain organisms (seeds and foods like yogurt that contain microorganisms are exceptions). It would likely be misleading to suggest that a food that ordinarily would not contain entire "organisms" is "organism free."

There is potential for the term "free" in a claim for absence of bioengineering to be inaccurate. Consumers assume that "free" of bioengineered material means that "zero" bioengineered material is present. Because of the potential for adventitious presence of bioengineered material, it may be necessary to conclude that the accuracy of the term "free" can only be ensured when there is a definition or threshold above which the term could not be used. FDA does not have information with which to establish a threshold level of bioengineered constituents or ingredients in foods for the statement "free of bioengineered material." FDA recognizes that there are analytical methods capable of detecting low levels of some bioengineered materials in some foods, but a threshold would require methods to test for a wide range of genetic changes at very low levels in a wide variety of foods. Such test methods are not available at this time. The agency suggests that the term "free" either not be used in bioengineering label statements or that it be in a context that makes clear that a zero level of bioengineered material is not implied. However, statements that the food or its ingredients, as appropriate, was not developed using bioengineering would avoid or minimize such implications. For example,

- "We do not use ingredients that were produced using biotechnology;"

- "This oil is made from soybeans that were not genetically engineered;" or

- "Our tomato growers do not plant seeds developed using biotechnology."

A statement that a food was not bioengineered or does not contain bioengineered ingredients may be misleading if it implies that the labeled food is superior to foods that are not so labeled. FDA has concluded that the use or absence of use of bioengineering in the production of a food or ingredient does not, in and of itself, mean that there is a material difference in the food. Therefore, a label statement that expresses or implies that a food is superior (e.g., safer or of higher quality) because it is not bioengineered would be misleading. The agency will evaluate the entire label and labeling in determining whether a label statement is in a context that implies that the food is superior.

In addition, a statement that an ingredient was not bioengineered could be misleading if there is another ingredient in the food that was bioengineered. The claim must not misrepresent the absence of bioengineered material. For example, on a product made largely of bioengineered corn flour and a small amount of soybean oil, a claim that the product "does not include genetically engineered soybean oil" could be misleading. Even if the statement is true, it is likely to be misleading if consumers believe that the entire product or a larger portion of it than is actually the case is free of bioengineered material. It may be necessary to carefully qualify the statement in order to ensure that consumers understand its significance.

Further, a statement may be misleading if it suggests that a food or ingredient itself is not bioengineered, when there are no marketed bioengineered varieties of that category of foods or ingredients. For example, it would be misleading to state "not produced through biotechnology" on the label of green beans, when there are no marketed bioengineered green beans. To not be misleading, the claim should be in a context that applies to the food type instead of the individual manufacturer's product. For example, the statement "green beans are not produced using biotechnology" would not imply that this manufacturer's product is different from other green beans.

Substantiation of label statements

A manufacturer who claims that a food or its ingredients, including foods such as raw agricultural commodities, is not bioengineered should be able to substantiate that the claim is truthful and not misleading. Validated testing, if available, is the most reliable way to identify bioengineered foods or food ingredients. For many foods, however, particularly for highly processed foods such as oils, it may be difficult to differentiate by validated analytical methods between bioengineered foods and food ingredients and those obtained using traditional breeding methods. Where tests have been validated and shown to be reliable they may be used. However, if validated test methods are not available or reliable because of the way foods are produced or processed, it may be important to document the source of such foods differently. Also, special handling may be appropriate to maintain segregation of bioengineered and non bioengineered foods. In addition, manufacturers should consider appropriate recordkeeping to document the segregation procedures to ensure that the food's labeling is not false or misleading. In some situations, certifications or affidavits from farmers, processors, and others in the food production and distribution chain may be adequate to document that foods are obtained from the use of traditional methods. A statement that a food is "free" of bioengineered material may be difficult to substantiate without testing. Because appropriately validated testing methods are not currently available for many foods, it is likely that it would be easier to document handling practices and procedures to substantiate a claim about how the food was processed than to substantiate a "free" claim.

FDA has been asked about the ability of organic foods to bear label statements to the effect that the food (or its ingredients) was not produced using biotechnology. On December 21, 2000, the Agriculture Marketing Service of the U.S. Department of Agriculture (USDA) published final regulations on procedures for organic food production (National Organic Program final rule; 65 FR 80548). That final rule requires that all but the smallest organic operations be certified by a USDA accredited agent and lays out the requirements for organic food production. Among those requirements is that products or ingredients identified as organic must not be produced using biotechnology methods. The national organic standards would provide for adequate segregation of the food throughout distribution to assure that non-organic foods do not become mixed with organic foods. The agency believes that the practices and record keeping that substantiate the "certified organic" statement would be sufficient to substantiate a claim that a food was not produced using bioengineering.

Adopted form FDA, CFSAN, Draft Guidance for Industry Regarding Labeling of Bioengineered Products, January 17, 2001

Section E: Botanical and Other Novel Ingredients in Conventional Foods

The Food and Drug Administration is concerned that some botanical and other novel ingredients that are being added to conventional foods are neither approved food additives nor generally recognized as being safe for these uses. Therefore, the agency issued a letter to the food industry restating the requirements of the Federal Food, Drug, and Cosmetic Act regarding the marketing of conventional foods containing novel ingredients, including botanicals. The issuance of this letter serves as a reminder to the industry of the longstanding legal requirements governing conventional food products. A copy of the letter is provided as "*Letter to Manufacturers*," for guidance purposes.

Letter to Manufacturers

Dear Manufacturer:

The Food and Drug Administration (FDA) has seen a significant growth in the marketplace of conventional food products that contain novel ingredients, such as added botanical ingredients or their extracts that have not previously been used as food ingredients. These products often bear claims to provide certain health benefits. FDA is writing to remind you about the requirements of the Federal Food, Drug, and Cosmetic Act (the Act) regarding the marketing of conventional foods containing novel ingredients, including botanicals. Of particular concern to the agency are the use of these ingredients and claims made on the label or in labeling.

Many ingredients intentionally added to a conventional food are food additives. Food additives require pre-market approval based on data demonstrating safety submitted to the agency in a food additive petition, ordinarily by the producer. The agency issues food additive regulations specifying the conditions under which an additive has been demonstrated to be safe and, therefore, may be lawfully used.

A substance is exempt from the definition of a food additive and thus, from pre-market approval, if, among other reasons, it is generally recognized as safe (GRAS) by qualified experts under the conditions of intended use. Accordingly, for a particular use of a substance to be GRAS, there must be both technical evidence of safety and a basis to conclude that this evidence is generally known and accepted by qualified experts. The technical element of the GRAS standard requires that the information about the substance establish that the intended use of the substance is safe, i.e., that there is a reasonable certainty in the minds of competent scientists that the substance is not harmful under its intended conditions of use. In addition, the data and information to establish the technical element must be generally available, and there must be a basis to conclude that there is consensus among qualified experts about the safety of the substance for its intended use. Any substance added to food that is an unapproved food additive (e.g., because it is not GRAS for its intended use) causes the food to be adulterated (Section 402(a)(2)(C) of the Act), and the food cannot be legally imported or marketed in the United States.

The FDA is concerned that some of the herbal and other botanical ingredients that are being added to conventional foods may cause the food to be adulterated because these added ingredients are not being used in accordance with an approved food additive regulation and may not be GRAS for their intended use.

Concerning label claims, the Act authorizes health claims, a claim characterizing the relationship between a substance and a disease or health-related condition, on food provided specific criteria are met following the submission of a petition and promulgation of a health claim regulation (Section 403(r)(1)(B)). In addition, the Act authorizes health claims based on authoritative statements through a notification procedure (Section 403(r)(3)(C)).

A food bearing a health claim that is not authorized by regulation or by the Act misbrands the product under section 403(r) of the Act. Currently, the health claims that FDA has authorized by regulation are listed in *21 C.F.R. 101.72 to 101.83* and include such claims as calcium and osteoporosis, sodium and hypertension, dietary lipids and cancer, folate and neural tube defects. As a legal matter, an unauthorized health claim or a claim that suggests that a food is intended to treat, cure or mitigate disease subjects the food to regulation as a drug (Section 201(g)(1)).

Food labels and labeling may also bear authorized nutrient content claims (Section 403(r)(1)(A)). A nutrient content claim is a claim characterizing the level of a nutrient in a food. As with health claims, FDA authorizes nutrient content claims following the submission of a petition and promulgation of a regulation; the Act also authorizes nutrient content claims based on authoritative statements through a notification procedure (Section 403(r)(2)(G)). A food bearing a nutrient content claim that has not been authorized by regulation or the Act misbrands the product (Section 403(r)(1)(A)). Currently, the nutrient content claims that FDA has authorized by regulation are located in *21 C.F.R. 101.13* and *21 C.F.R. 101.54 to 101.67*. Some nutrient content claims, such as "high" and "more" are defined only for substances with an established Reference Daily Intake (RDI) or Daily Reference Value (DRV). A list of nutrients with RDIs can be found at *21 C.F.R. 101.9(c)(8)(iv)*; nutrients with DRVs are listed in *21 C.F.R. 101.9(c)(9)*. Other nutrient content claims may be made for any substance, provided that the requirements of the authorizing regulation are met. For example, a manufacturer may make a statement about a substance for which there is no established RDI or DRV so long as the claim specifies only the amount of the substance per serving and does not imply that there is a lot or a little of the substance in the product.

A conventional food label or labeling may bear statements about a substance's effect on the structure or function of the body provided such statements do not claim to diagnose, mitigate, treat, cure, or prevent disease and are not false or misleading. Additionally, the claimed effect must be achieved through nutritive value.. In the preamble to the final rule governing structure/function claims on dietary supplements (65 FR 1000 at 1034; January 6, 2000), FDA stated that the agency is likely to interpret the dividing line between structure/function claims and disease claims in a similar manner for conventional foods as for dietary supplements. As a legal matter, a structure/function claim on a food that is not achieved through nutritive value, or a claim that a food will treat or mitigate disease, subjects the product to regulation as a drug under section 201(g)(1) of the Act.

Questions regarding the regulatory status of ingredients that you intend to use in your conventional foods and how to file a GRAS Notification or Food Additive Petition should be directed to Dr. George Pauli, Director of the Division of Product Policy, Office of Premarket Approval, Center for Food Safety and Applied Nutrition, HFS-205, 200 C Street, S.W., Washington, D.C. 20204. Questions regarding labeling claims for these foods should be directed to Mr. John B. Foret, Director of the Division of Compliance and Enforcement, Office of Nutritional Products, Labeling and Dietary Supplements, Center for Food Safety and Applied Nutrition, HFS-810, 200 C Street, S.W., Washington, D.C. 20204.

FDA's general food labeling requirements are located in Title 21 of the *Code of Federal Regulations* part 101, and additional guidance can be obtained from the <u>Food Labeling Guide</u> that is available on the FDA Web page, www.fda.gov.

Chapter X

Sincerely yours,

Christine J. Lewis, Ph.D.
Director
Office of Nutritional Products, Labeling and Dietary Supplements
Center for Food Safety and Applied Nutrition

Adopted from, U.S. FDA, FDA Talk Paper, "FDA Issued Letter To Industry On Foods Containing Botanical And Other Novel Ingredients," 2/5/01

Section F: Guidance for Industry:

Guidance on the Labeling of Certain Uses of Lecithin Derived from Soy under Section 403(w) of the Federal Food, Drug, and Cosmetic Act

Comments and suggestions regarding this document may be submitted at any time. Submit comments to Division of Dockets Management (HFA-305), Food and Drug Administration, 5630 Fishers Lane, rm. 1061, Rockville, MD 20852. All comments should be identified with the docket number listed in the notice of availability that publishes in the *Federal Register*.

For questions regarding this document contact Paul M. Kuznesof, Ph.D., at the Center for Food Safety and Applied Nutrition (CFSAN) at 301-436-1289.

Guidance

Consistent with the need to establish its enforcement priorities, FDA intends to consider the exercise of enforcement discretion for a food labeled on or after January 1, 2006, in which lecithin derived from soy is used as a component of a release agent and the label for such food does not declare the presence of lecithin consistent with the requirements of section 403(w) of the Act. The agency's intent to exercise its enforcement discretion for a limited period for the foregoing use of lecithin will help FDA to apply its increasingly limited resources to efforts associated with implementation and enforcement of FALCPA that are expected to have a more acute public health impact.

The agency intends to reconsider its enforcement priorities with regard to the labeling of lecithin derived from soy used as a component of a release agent approximately 18 months after the issuance of this guidance. The agency expects that, during the period in which FDA intends to consider the exercise of its enforcement discretion as described above, manufacturers of foods that use lecithin derived from soy as a component of a release agent will revise as necessary the labels of their relevant food products to comply with FALCPA and begin to label their products using the FALCPA-compliant labels by the end of the enforcement discretion period.

FDA intends to consider exercising such discretion when all of the following factors are present:

1. The food was labeled on or after January 1, 2006.
2. The lecithin derived from soy used as a component of a release agent satisfies each of the specifications for lecithin in the *Food Chemicals Codex*, 5th Edition.
3. The lecithin derived from soy is used solely as a component of a release agent as described in this guidance.
4. The release agent in which lecithin derived from soy is a component is used at the lowest level possible consistent with current good manufacturing practice.

The agency emphasizes that this guidance does not apply if the lecithin derived from soy used as a component of a release agent does not comply with the *Food Chemicals Codex* specifications or if the lecithin derived from soy is used other than as a component of a release agent as described in this guidance.

Copies of this document are available from the Office of Food Additive Safety (HFS-205), Center for Food Safety and Applied Nutrition, Food and Drug Administration, 5100 Paint Branch Parkway, College Park, MD 20740

Adopted from FDA, CFSAN/Office of Food Additive Safety, Guidance for Industry Guidance on the Labeling of Certain Uses of Lecithin Derived from Soy Under Section 403(w) of the Federal Food, Drug, and Cosmetic Act, April 2006

Section G. Guidance for Industry:

Food Labeling: Safe Handling Statements, Labeling of Shell Eggs; Refrigeration of Shell Eggs Held for Retail Distribution (Small Entity Compliance Guide)

The Food and Drug Administration has prepared this Small Entity Compliance Guide in accordance with section 212 of the Small Business Regulatory Fairness Act (P.L. 104-121). This guidance document restates in plain language the legal requirements set forth in the current regulations for the safe handling statement on labels of shell eggs and the refrigeration of shell eggs held at retail establishments. This is a Level 2 guidance document published for immediate implementation in accordance with FDA's good guidance practices *(21 CFR 10.115)*. The regulations are binding and have the force and effect of law. However, this guidance document represents the agency's current thinking on this subject and does not, itself, create or confer any rights for or on any person and does not operate to bind FDA or the public. An alternative approach may be used if such approach satisfies the requirements of the applicable statute and regulations.

SUMMARY

The Food and Drug Administration (FDA) published a final rule in the *Federal Register* of December 5, 2000 (65 FR 76092) entitled, "Food Labeling: Safe Handling Statements, Labeling of Shell Eggs; Refrigeration of Shell Eggs Held for Retail Distribution." The final rule applies to shell eggs that have not been specifically processed to destroy all live *Salmonellae* before distribution to the consumer. For these shell eggs, retail establishments must include the following safe handling statement on the label of the shell eggs:

SAFE HANDLING INSTRUCTIONS:

To prevent illness from bacteria: keep eggs refrigerated, cook eggs until yolks are firm, and cook foods containing eggs thoroughly.

The regulation also requires retail establishments to refrigerate shell eggs promptly when they are received and to store the eggs at 45° F (7.2° C) or cooler.

FDA is requiring these actions to reduce the risk of illness and death caused by *Salmonella* Enteritidis (SE), a pathogenic bacterium, which is associated with the consumption of shell eggs that have not been treated to destroy *Salmonella*.

QUESTIONS AND ANSWERS

GENERAL INFORMATION

What products are subject to the requirements of this regulation?

Answer: All shell eggs that have not been treated specifically to kill all live *Salmonellae* before distribution to the consumer are subject to the requirements of this regulation.

Are eggs sold in intrastate commerce (i.e., eggs sold only in the state where they were produced) subject to this regulation?

Chapter X

Answer: Yes. Eggs sold in both intrastate and interstate commerce (i.e., sold across state or international lines) are required to comply with this regulation.

Are egg products (i.e., liquid, dried, and frozen whole eggs, egg whites, and egg yolks, with or without added ingredients) subject to this regulation?

Answer: No. FDA already has a separate regulation for egg products as stated in the standards of identity for egg products *(21 CFR part 160)*, which requires those products to be pasteurized or otherwise treated to destroy all live *Salmonella*. Therefore, egg products are not subject to this regulation. However, FDA recommends that those products be kept refrigerated or frozen to ensure quality. Dried egg products may be stored at room temperature.

Are in-shell pasteurized eggs subject to this regulation?

Answer: No. Eggs that have been pasteurized while in the shell have been treated specifically to kill all live *Salmonella* and, therefore, are not subject to this regulation. However, FDA recommends that in-shell pasteurized eggs be refrigerated to retain shelf life.

Are shell eggs produced under Quality Assurance Programs (QAP's) subject to this regulation?

Answer: Yes. Shell eggs produced under a QAP are subject to this regulation.

SAFE HANDLING STATEMENT AND LABELING REQUIREMENT

What safe handling statement must be included in the label of shell eggs that have not been treated specifically to kill all live Salmonella before the shell eggs may be distributed to the consumer?

Answer: The safe handling statement is: "SAFE HANDLING INSTRUCTIONS: To prevent illness from bacteria: keep eggs refrigerated, cook eggs until yolks are firm, and cook foods containing eggs thoroughly."

How must the safe handling statement appear on the label?

Answer: The statement must appear on the label prominently, conspicuously, and in a type size no smaller than one-sixteenth of one inch. The statement must appear in a hairline box and the words "safe handling instructions" must appear in bold capital letters. For example:

> SAFE HANDLING INSTRUCTIONS: To prevent illness from bacteria: keep eggs refrigerated, cook eggs until yolks are firm, and cook foods containing eggs thoroughly.

For shell eggs that are sold to consumers, where must the safe handling statement appear on the label?

Answer: The statement must appear on either: (1) the principal display panel (PDP) (i.e., that part of the label most likely to be seen by the consumer at the time of purchase); or (2) the information panel (i.e., the panel immediately to the right of the PDP or, in the instance where the top of the package is the PDP, any panel adjacent to the PDP).

Must the safe handling statement appear on the package of shell eggs that are not for direct sale to consumers but rather are to be repackaged, relabeled or further processed (e.g., cases of shell eggs sold to food service establishments)?

Answer: No. Under such circumstances, there are no placement requirements for the safe handling statement. You may choose to place the safe handling instructions on the package label. Alternatively, you may choose to place the safe handling statement in the labeling (i.e., invoices and bills of lading), according to trade practices. However, if you choose not to place FDA's required statement on the label, you must comply with USDA's labeling requirements, specifically USDA's "Keep Refrigerated" labeling requirement, which can be found at *9 CFR § 590.50.*

By what date must the safe handling statement appear on the label?

Answer: The safe handling statement must appear on the label of shell eggs by September 4, 2001. However, FDA encourages you to begin voluntarily labeling your products now.

Can eggs, currently labeled with a safe handling statement that is different from the one FDA is requiring, continue to bear that existing statement?

Answer: Eggs that currently have a different safe handling statement on their labels may continue to do so until the supply of labels is used up, <u>only</u> if the statement includes a refrigeration instruction and a cooking instruction. Once the supply of labels is used up, you must comply with FDA's prescribed statement.

REFRIGERATION REQUIREMENT

Which retail establishments are subject to the refrigeration provisions of this regulation?

Answer: Any retail establishment that is an operation that stores, prepares, packages, serves, sells, or otherwise provides food for human consumption directly to consumers is subject to the refrigeration provisions of this regulation. This includes, but is not limited to, supermarkets, restaurants, delicatessens, caterers, vending operations, hospitals, nursing homes, and schools.

What are retail establishments required to do about refrigerating shell eggs that have not been specifically treated to kill all live Salmonellae?

Answer: Upon receipt, retail establishments promptly must store and display the eggs at an ambient (air) temperature of 45° F (7.2° C) or cooler.

How soon after receiving eggs must a retail establishment refrigerate them?

Answer: Retail establishments must refrigerate the eggs promptly upon receipt, except when short delays are unavoidable. In these situations, the eggs must be refrigerated as soon as reasonably possible.

By what date must retail establishments comply with the refrigeration requirements for shell eggs?
Answer: Retail establishments already must be complying with the refrigeration provisions of this regulation since they were effective on June 4, 2001.

Adopted from U. S. FDA, CFSAN, ONPLDS OPDFB," Guidance for Industry Food Labeling: Safe Handling Statements, Labeling of Shell Eggs; Refrigeration of Shell Eggs Held for Retail Distribution, July 2001

Chapter X

Section H. Guidance on Labeling of Foods That Need Refrigeration by Consumers

ACTION: Notice.

SUMMARY: The Food and Drug Administration (FDA) is providing guidance on labeling of foods that need refrigeration by consumers to maintain safety or quality. This guidance, which represents FDA's policy on adequate safe handling instructions for food, should reduce the likelihood of temperature abuse of certain foods by consumers, and it is intended to reduce the potential for foodborne illness and death.

The guidance also responds to the recommendations of the National Advisory Committee on Microbiological Criteria for Foods (NACMCF), the National Food Processors Association (NFPA), the Association of Food and Drug Officials (AFDO), and the Centers for Disease Control and Prevention (CDC) for labeling foods needing refrigeration. FDA is soliciting comments on this guidance.

DATES: Written comments may be submitted at any time.

ADDRESSES: Submit written comments on this guidance to the Dockets Management Branch (HFA-305), Food and Drug Administration, 12420, Parklawn Dr., rm. 1-23, Rockville, MD 20857.

FOR FURTHER INFORMATION CONTACT: Geraldine A. June, Center for Food Safety and Applied Nutrition (HFS-158), Food and Drug Administration, 200 C St. SW., Washington, DC 20204, 202-205-5099.

FDA Labeling Policy

To clarify this guidance, the agency has delineated each of the three groups and developed model statements for each:

1. Group A Foods

Group A foods are potentially hazardous foods, which, if subjected to temperature abuse, will support the growth of infectious or toxigenic microorganisms that may be present. Outgrowth of these microorganisms would render the food unsafe. Foods that must be refrigerated for food safety possess the following characteristics: (1) Product pH > 4.6; (2) water activity a_w > 0.85; (3) do not receive a thermal process or other treatment in the final package that is adequate to destroy foodborne pathogens that can grow under conditions of temperature abuse during storage and distribution; and (4) have no barriers (e.g., preservatives such as benzoates, salt, acidification), built into the product formulation that prevent the growth of foodborne pathogens that can grow under conditions of temperature abuse during storage and distribution.

The appropriate label statement for Group A foods is:

> IMPORTANT: Must Be Kept Refrigerated To Maintain Safety

2. Group B Foods

Group B includes those foods that are shelf-stable as a result of processing, but once opened, the unused portion is potentially hazardous unless refrigerated. These foods possess the following characteristics: (1) Product pH > 4.6; (2) water activity a_w > 0.85; (3) receive a thermal process or other treatment that is adequate to destroy or inactivate foodborne pathogens in the unopened package, but after opening, surviving or contaminating microorganisms can grow and render the product unsafe; and (4) have no barriers (for example, preservatives such as benzoates, salt, acidification) built into the product formulation to prevent the growth of foodborne pathogens after opening and subsequent storage under temperature abuse conditions. The appropriate label statement for Group B foods is:

> IMPORTANT: Must Be Refrigerated After Opening To Maintain Safety

3. Group C Foods

Group C are those foods that do not pose a safety hazard even after opening if temperature abused, but that may experience a more rapid deterioration in quality over time if not refrigerated. The manufacturer determines whether to include on the label a statement that refrigeration is needed to maintain the quality characteristics of the product to maximize acceptance by the consumer. These foods do not pose a safety problem. Foods in this group possess one or more of the following characteristics to ensure that the food does not present a hazard if temperature abused: (1) Product pH <ls-thn-eq> 4.6 to inhibit the outgrowth and toxin production of C. botulinum; or (2) water activity a<INF>w <ls-thn-eq> 0.85; or (3) have barriers built into the formulation (for example, preservative systems such as benzoates, salt, acidification) to prevent the growth of foodborne pathogens if the product is temperature abused.

The suggested optional label statement for Group C foods is:

"Refrigerate for Quality" or some other statement that explains to the consumer that the storage conditions are recommended to protect the quality of the product. To avoid confusion between refrigeration for safety purposes and refrigeration for quality reasons, Group A and Group B statements should not be used on Group C foods.

The agency is publishing this document to provide this guidance by the quickest means to as many manufacturers as possible, so that they may begin using the label statements. If manufacturers follow this guidance, the consumer will have clear, concise, and prominent labeling information for maintaining the safety of potentially hazardous food products. Inclusion of these statements in the labeling of appropriate foods will help the consumer recognize when appropriate storage temperatures are needed to maintain the safety or quality of those foods. Such information will reduce the likelihood of temperature abuse of the food and, consequently, reduce the potential for foodborne illness and death. While this guidance is primarily intended to address the need for safe handling of potentially hazardous foods by consumers, the agency recognizes that there also is a need for safe handling during the transportation and distribution of these foods. The Food Safety and Inspection Service of the U.S. Department of Agriculture and FDA have jointly published an advance notice of proposed rulemaking in the *Federal Register* of November 22, 1996 (61 FR 59372) to solicit comments on approaches that the two agencies may take to foster safety improvements in the storage and transportation of potentially hazardous foods. Therefore, this guidance does not address how foods that need refrigeration during transportation and storage should be labeled.

Adopted from the Federal Register Online via GPO Access (wais.access.gpo.gov) DOCID: fr24fe97-74, Notice "Guidance on Labeling of Foods That Need Refrigeration by Consumers", February 24, 1997 (Volume 62, Number 36), Page 8248-8252

Section I. Food Labeling; Gluten-Free Labeling of Foods; Public Meeting

ACTION: Notice of public meeting; request for comments.

SUMMARY: The Food and Drug Administration (FDA) is announcing a public meeting to obtain expert comment and consultation from stakeholders to help the agency to define and permit the voluntary use on food labeling of the term "gluten-free". The meeting focused on food manufacturing, analytical methods, and consumer issues related to reduced levels of gluten in food.

I. Background

Celiac disease (also known as celiac sprue) is a chronic inflammatory disorder of the small intestine triggered by ingesting certain storage proteins that naturally occur in cereal grains. Celiac disease is genetically inherited, and its prevalence in the United States is estimated to be slightly less than 1 percent of the general population.

The grains that are considered to cause problems for persons with celiac disease are wheat, barley, and rye, their related species (e.g., durum wheat, spelt, kamut) and crossbred hybrids (e.g., triticale), and possibly oats. The scientific literature includes reports of celiac disease patients who can tolerate oats and others who cannot. This intolerance may be due to the possible presence in commercially available oat products of trace amounts of other grains that are harmful to persons who have celiac disease (e.g., wheat, rye, or barley). However, there is also some evidence that naturally occurring proteins in uncontaminated oats may cause adverse effects in some celiac disease patients.

Technically, the term "gluten" applies to the combination of storage proteins found in wheat, the prolamin proteins called "gliadins" and the glutelin proteins called "glutenins". However, in the context of celiac disease, the term "gluten" is often used to refer collectively to any of the proteins in the grains that may cause harm. Currently, to prevent severe and sometimes life-threatening complications of celiac disease, sensitive individuals need to avoid all offending sources of gluten. Life-threatening complications can affect multiple organs of the body.

The Food Allergen Labeling and Consumer Protection Act of 2004 (FALCPA) (Title II of Public Law 108-282) at http://www.cfsan.fda.gov/~dms/alrgact.html requires FDA to issue, within 2 years of the enactment date, a proposed rule to define, and permit the use of, the term `gluten-free" on food labeling and a final rule within 4 years of enactment. FALCPA requires FDA to consult with appropriate experts and stakeholders during the agency's development of the proposed rule. Establishing a definition of 'gluten-free" that is both protective of the celiac population and that uniformly applies to "gluten-free" labeling statements for foods marketed in the United States will assist Americans with celiac disease to make more informed food consumption decisions.

II. Purpose and Scope of Meeting

FDA is holding this meeting to solicit comments from appropriate experts and stakeholders to assist us in developing a proposed rule to define and permit the use of the term ``gluten-free," as required by FALCPA. The agency is interested in gathering information from the public, particularly the food industry on how "gluten-free" foods are manufactured, the analytical methods used to verify that foods are "gluten-free," and related costs of manufacturing "gluten-free" foods. The agency is also interested in receiving research data or findings on the food purchasing practices of consumers with celiac disease and their caregivers related to packaged products labeled or marketed as ``gluten-free," compared to their purchasing practices of packaged products that are not so labeled. The public meeting will not address issues regarding a threshold level of gluten (i.e., the amount of gluten below which it would be unlikely to elicit harmful effects in celiac disease patients) and the medical implications of celiac disease. These two issues were addressed at a meeting of FDA's Food Advisory Committee (FAC) on July 13 through 15, 2005 (70 FR 29528, May 23, 2005). The meeting agenda provided that the FAC would review and evaluate the Center for Food Safety and Applied Nutrition Threshold Working Group draft report entitled ``Approaches to Establish Thresholds for Major Food Allergens and for Gluten in Food," which may be found on the Internet at http://www.cfsan.fda.gov/~dms/alrgn.html.

FDA will consider all pertinent information, including the recommendations of the FAC and comments from this public meeting, in developing a definition and establishing the permissible use of the term "gluten-free" in food labeling.

III. Questions

FDA has drafted a series of questions to help focus the comments presented at the public meeting or otherwise communicated to the agency. Those who comment are invited to address any or all of these questions. FDA is particularly interested in receiving related technical, scientific, and cost data from the food industry as well as research data or findings about the food purchasing practices of consumers with celiac disease or their caregivers. For the purpose of the list of questions in this document, FDA is using the following terms:

"Gluten" refers to the proteins found in any of the grains that can cause harm to persons with celiac disease;

"Grains of concern" refers to wheat, rye, barley, and oats, and their related species (e.g., durum, spelt, kamut) or crossbred hybrids (e.g., triticale); and

"Gluten-free foods" refers to foods currently marketed in the United States that are either represented to be free of gluten or that contain statements or symbols on their labeling that identify the products as ones that do not contain gluten.

A. Definitions of "Gluten-Free"

1. How do food manufacturers define "gluten-free"? What is the generally accepted definition in the food industry of "gluten-free"?

Please identify any entities that "certify" finished foods or raw ingredients to be "gluten-free". Describe how they define "gluten-free" and how they determine whether a food product satisfies this definition.

B. "Gluten-Free" Product Development

2. How are "gluten-free" foods produced? For example, are "gluten-free" foods made by using only ingredients that do not contain any gluten (i.e., they are inherently "gluten-free") or are they made by processing ingredients or the finished food to remove gluten? What methods are most commonly used to remove gluten from food?
3. Due to potential grain cross-contact situations, is it technologically feasible to produce "gluten-free" flour from grains other than those of concern (e.g., corn, millet)? Is it technologically feasible to produce oat-based products that do not contain gluten from grains of concern other than oats (e.g., wheat)? If so, what additional measures in the milling or manufacturing process would be needed to produce these products? Is it economically feasible to produce such products, and if so, what would be the incremental costs?

C. Good Manufacturing Practices and Analytical Methods

4. What measures do you have in place during the manufacturing, packaging, or holding of "gluten-free" foods to prevent them from coming into contact with any grains of concern? For example, do you use dedicated facilities, dedicated equipment, or dedicated production lines?
5. What analytical method(s) do you use to evaluate your "gluten-free" products? How often do you perform these analyses? For example, do you test every batch of finished product? Do you test bulk containers of each ingredient? What is the cost of such testing?
6. The following questions seek data and information about available gluten detection test kits or analytical methods to detect gluten:

- In what grains can the test kit or method detect gluten?
- What specific mechanism is used to indicate the presence or absence of gluten?
- What is the sensitivity or lowest level of detection of your test kit or method?
- Is your test kit or method qualitative (i.e., establishes only the presence or absence of gluten) or quantitative?
- If quantitative, what is the limit of quantification of your test kit or method?
- What is the false positive rate of your test kit or method? What is its false negative rate?
- Is the effectiveness of your test kit or method affected by the nature of the processing of the ``gluten-free" food, and if so, how? Is it affected by the food matrix, and if so, how? (FDA is especially interested in information that addresses the influence of the presence of fermented or hydrolyzed proteins, of xanthan gum, of guar gum, or of any other dietary fibers.)
- If your test kit or method has been validated, please indicate by whom it was validated and the level (e.g., parts per million) of detection at which it was validated.
- If your test kit or method has not been validated, have the results of its performance or an evaluation of its performance been published in a peer-reviewed scientific journal?
- What is the cost of your test kit or the cost to perform your method of analysis?

Chapter X

7. What analytical methods are currently available or under development to detect the presence of oat proteins in food? Please specify which proteins. What is the cost to conduct such analyses? Have any of these methods been validated or published in a peer-reviewed scientific journal?

D. Foods Marketed as "Gluten-Free"

8. Are there available research data or findings on what consumers with celiac disease or their caregivers believe the term "gluten-free" means? For example, do the research data or findings show consumers' beliefs as to which specific grains or other ingredients are not present in foods labeled "gluten-free"?

E. Consumer Purchasing Practices

9. Are there available research data or findings on how consumers with celiac disease or their caregivers identify packaged foods that do not contain gluten? Do the data establish how much time these consumers devote to identifying such foods?

10. Are there available research data or findings on whether the packaged foods consumers with celiac disease or their caregivers currently purchase or consume are primarily or exclusively those foods labeled "gluten-free"? Do the research data or findings identify the types of "gluten-free" packaged foods (e.g., breads, dairy foods, canned vegetables) purchased or consumed by persons with celiac disease or their caregivers? Do the research data or findings show whether a "gluten-free" label influences the purchasing decision of persons with celiac disease or their caregivers when presented with products having identical ingredient lists?

A transcript will be made of the meeting's proceedings. You may request a copy in writing from FDA's Freedom of Information Office (HFI-35), Food and Drug Administration, 5600 Fishers Lane, rm. 12A-16, Rockville, MD 20857, approximately 30 working days after the public meeting at a cost of 10 cents per page. The transcript of public meeting and all comments submitted will be available for public examination at the Division of Dockets Management between 9 a.m. and 4 p.m., Monday through Friday, as well as on the FDA Web site at http://www.fda.gov/ohrms/dockets/default.htm.

Adopted From the Federal Register July 19, 2005 (Volume 70, Number 137) online via GPO Access [wais.access.gpo.gov] [DOCID:fr19jy05-12], Proposed Rules "Food Labeling; Gluten-Free Labeling of Foods; Public Meeting; Request for Comments", [Page 41356-41358]

Section J. Guidance for Industry and FDA Staff: Whole Grain Label Statements; Draft Guidance

(This guidance document is being distributed for comment purposes only).

Comments and suggestions regarding this draft document should be submitted within 60 days of publication in the *Federal Register* of the notice announcing the availability of the draft guidance. Submit comments to the Division of Dockets Management (HFA-305), Food and Drug Administration, 5630 Fishers Lane, Rm. 1061, Rockville, MD 20852. All comments should be identified with the docket number listed in the notice of availability that publishes in the *Federal Register*.

This draft guidance, when finalized, will represent the Food and Drug Administration's (FDA's) current thinking on this topic. It does not create or confer any rights for or on any person and does not operate to bind FDA or the public. You can use an alternative approach if the approach satisfies the requirements of the applicable statutes and regulations. If you want to discuss an alternative approach, contact the FDA staff responsible for implementing this guidance. If you cannot identify the appropriate FDA staff, call the appropriate number listed on the title page of this guidance.

1. Introduction

This guidance is intended for the regulated food industry and FDA personnel. The purpose of this guidance is to provide guidance to industry about what the agency considers to be "whole grain" and to assist manufacturers in labeling their products.

FDA's guidance documents, including this guidance, do not establish legally enforceable responsibilities. Instead, guidances describe the Agency's current thinking on a topic and should be viewed only as recommendations, unless specific regulatory or statutory requirements are cited. The use of the word *should* in Agency guidances means that something is suggested or recommended, but not required.

2. Background

Through the years, the Federal Government has worked to provide consistent and scientifically sound recommendations to consumers about healthy eating patterns and wise food choices. Such advice originated with the "Basic Four" and has progressed through today's "Dietary Guidelines for Americans" (developed jointly by the U.S. Department of Health and Human Services and the U.S. Department of Agriculture (USDA)). "Dietary Guidelines for Americans, 2005" (2005 DG) recommends that Americans, among other things, "consume 3 or more ounce-equivalents of whole grain products per day, with the rest of the recommended grains coming from enriched or whole-grain products" and that "in general at least half the grains should come from whole grains".

Manufacturers can make factual statements about whole grains on the label of their products such as "100% whole grain " (as percentage labeling under *21 CFR 102.5(b)*) or "10 grams of whole grains " *(21 CFR 101.13(i)(3))* provided that the statements are not false or misleading under section 403(a) of the Federal Food, Drug, and Cosmetic Act (the Act) and do not imply a particular level of the ingredient, i.e., "high " or "excellent source. " In addition, manufacturers may use health claims relating whole grains with a reduced risk of coronary heart disease and certain cancers on their product labels for qualifying foods based on notifications FDA received under section *403(r)(3)(C)* of the Act *(21 U.S.C. 343 (r)(3)(C))* (health claims based on an authoritative statement of a scientific body) (see FDA Modernization Act of 1997 (FDAMA) Claims). To assist manufacturers in labeling their products in accordance with the Act, the agency has reviewed various industry and scientific definitions of "whole grains" and developed the following questions and answers to provide guidance to industry about what the agency considers to be "whole grain."

3. Questions and Answers

What factors should be considered in determining whether a food is a whole grain?

Answer: Cereal grains that consist of the intact, ground, cracked or flaked caryopsis, whose principal anatomical components - the starchy endosperm, germ and bran - are present in the same relative proportions as they exist in the intact caryopsis - should be considered a whole grain food.

What are some examples of cereal grains?

Answer: Cereal grains may include amaranth, barley, buckwheat, bulgur, corn (including popcorn), millet, quinoa, rice, rye, oats, sorghum, teff, triticale, wheat, and wild rice.

Should soybeans and chickpeas be considered whole grains?

Answer: Soybeans and chickpeas should not be considered whole grains, but should be considered legumes. Products derived from legumes, oilseeds (sunflower seeds), and roots (e.g., arrowroot) should not be considered whole grains.

Should a corn flour or corn meal made from corn grain to which the pericarp has been removed be considered whole grain?

Answer: The four principal parts of a mature corn kernel consist of the hull or bran (pericarp and seed coat), germ, endosperm, and the tip cap. The tip cap, the attachment point of the cob, may or may not stay with the kernel during handling, and, thus, is not considered an integral part of the kernel or caryopsis. However, the bran, germ and endosperm are integral parts of the kernel and should be present in the relative proportions as found in the kernel to be considered "whole grain." Therefore, for corn flour or corn meal to be "whole grain" it should include the pericarp as well as the other essential fractions.

Chapter X

We note that there are standards of identity for various types of corn flour and corn meal in *21 CFR Part 137* (i.e., *§ 137.211*, white corn flour; *§ 137.215*, yellow corn flour; *§ 137.250*, white corn meal; *§ 137.255*, bolted white corn meal; *§ 137.260*, enriched corn meals; *§ 137.265*, degerminated white corn meal; *§ 137.270*, self-rising white corn meal; *§ 137.275*, yellow corn meal; *§ 137.280*, bolted yellow corn meal; *§ 137.285*, degerminated yellow corn meal; and *§ 137.290*, self-rising yellow corn meal). Degerminated and bolted corn meals should not be considered whole grain products because germ or bran has been removed during processing. Because the rest of the meal standards allow removal of some of the hull, these also should not be considered whole grain products.

Barley has a particularly tough hull and is often pearled to make it easier to cook and digest. Can the hull and perhaps a small amount of the bran attached to the hull be removed from barley in the pearling process and it still be considered a whole grain?

Answer: Most of the barley that is used for food production in the U.S. is of a type in which the kernels are covered with a very tough inedible hull. This outer hull (which covers the bran layer) must be removed before the kernel can be used for human food. The hull on many varieties of barley is strongly attached to the pericarp. Thus, barley is difficult to dehull and generally is pearled. The pearling process abrades away the outer surfaces of the grain with an abrasive surface and removes some of the bran from the barley.

In general, the barley that is used for human food in the U.S. is pearled. Barley that is pearled should not be considered a whole grain because some of the bran layer has been removed. Dehulled barley should be considered a whole grain because only the tough inedible hull or outer covering has been removed, but the bran layer is left intact.

Should rolled oats be considered a whole grain?

Answer: In the U.S. most oats are flattened to produce rolled oats, or steamed and flattened to create "quick oats." Rolled oats and "quick oats" processed simply by flattening and/or steaming should be considered whole grains because they contain all of the bran, germ, and endosperm of whole oats.

Does the term "whole grain" mean the same as "100 percent whole grain"? If a product is labeled as "whole wheat bagel" or "whole wheat pizza," how much whole wheat should it contain? What is graham flour?

Answer: FDA has not defined any claims concerning the grain content of foods. However, the agency has established standards of identity for various types of cereal flours and related products in *21 CFR Part 137*, including a standard of identity for "whole wheat flour" *(§ 137.200)* and "whole durum flour" *(§ 137.225)*. Graham flour is an alternative name for whole wheat flour *(§ 137.200)*.

Depending on the context in which a "whole grain" statement appears on the label, it could be construed as meaning that the product is "100 percent whole grain." We recommend that products labeled with "100 percent whole grain" not contain grain ingredients other than those the agency considers to be whole grains. Consumers should be able to look at the ingredient statement to determine whether the predominant or first ingredient listed is a whole grain. We note that wheat flour should not be labeled as a whole grain flour because wheat flour is a synonym of flour *(§ 137.105)*, and thus, the bran and germ have been removed. However, whole wheat flour *(§ 137.200)* should be considered a whole grain flour because it contains all the parts of the grain, i.e., the bran, endosperm, and germ. We recommend that pizza that is labeled "whole grain" or "whole wheat" only be labeled as such when the flour ingredient in the crust is made entirely from whole grain flours or whole wheat flour, respectively. Similarly, we recommend that bagels, labeled as "whole grain " or "whole wheat" only be labeled as such when bagels are made entirely from whole grain flours or whole wheat flour, respectively.

What is durum wheat? Is it 100 percent whole grain? What products are made from durum wheat?

Answer: Durum wheat is a type of wheat that has a high protein content and the flour has a yellow color. It is typically used for semolina and pastas. Durum flour should not be considered a whole grain flour because the germ and bran have been removed *(21 CFR 137.220)*. However, whole durum flour *(21 CFR*

137.225) should be considered a whole grain flour because the flour contains all the parts of the grain, i.e., the bran, endosperm, and germ. We recommend that products labeled with "100 percent durum wheat" statements be made entirely with durum flour and products labeled "whole grain" be made entirely from whole durum flour.

Are there standards of identity for products made from whole grains?

Answer: There are no standards of identity for whole grain products per se. However, there are standards of identity for whole wheat bread, rolls, and buns *(21 CFR 136.180)* and whole wheat macaroni products *(21 CFR 139.138)* which are made from whole wheat flours. For bread, rolls, and buns, the dough is made from whole wheat flour, brominated whole wheat flour, or a combination of these and no other type of flour is used. Whole wheat macaroni products are made from whole wheat flour, whole durum wheat flour, or both.

What types of label statements about whole grains are currently permitted to be made on food products?

Answer: Manufacturers can make factual statements about whole grains on the label of their products, such as "10 grams of whole grains," "½ ounce of whole grains," *(21 CFR 101.13(i)(3))* and "100% whole grain oatmeal" (as percentage labeling under *21 CFR 102.5(b)*), provided that the statements are not false or misleading under section 403(a) of the Act and do not imply a particular level of the ingredient, i.e., "high" or "excellent source."

In addition, labels may bear a health claim based on an authoritative statement of a scientific body relating whole grains with a reduced risk of heart disease and certain cancers if the food meets the qualifications of one of the notifications submitted under section 403(r)(3)(C) of the Act (see FDA Modernization Act of 1997 (FDAMA) Claims).

Can the name of the particular whole grain be substituted for the term "whole grain" in label statements? For example, could the statement "100% brown rice" replace the statement "100% whole grains" or "1 ounce whole wheat " replace "1 ounce whole grain?"

Answer: The specific name of the whole grain (e.g., brown rice) can be used for label statements made under *21 CFR 102.5(b)* or *21 CFR 101.13(i) (3)* as long as the statement is truthful and not misleading. However, "whole grains" is the substance of the health claims established under section 403(r) (3)(C) of the Act and the name of a particular whole grain can not be substituted for the term "whole grain foods" in the health claims.

For questions regarding this draft document contact Shellee Anderson, Food Labeling and Standards Staff (HFS-820), Office of Nutritional Products, Labeling and Dietary Supplements, Center for Food Safety and Applied Nutrition, 5100 Paint Branch Parkway, College Park, Maryland 20740, (301) 436-2371.

Copies of the complete report are available from Food Labeling and Standards Staff (HFS-820), Office of Nutritional Products, Labeling and Dietary Supplements, Center for Food Safety and Applied Nutrition, Food and Drug Administration, 5100 Paint Branch Parkway, College Park, MD 20740, (Tel) 301-436-2371, http://www.fda.gov/ohrms/dockets/default.htm

Adopted from FDA, CFSAN, Guidance for Industry and FDA Staff: Whole Grain Label Statements, February 15, 2006

Chapter XI
Charts/Illustrations/Statements/ Regulations

(See the Specific Page Number for the Desired Attachment)

Chapter XI

Chapter XII

Index to the January 6, 1993 Federal Register Preamble and the Final NLEA Regulations

---

INDEX TO THE JANUARY 6, 1993 *FEDERAL REGISTER* PREAMBLE AND THE FINAL NLEA REGULATIONS

FOOD LABELING REGULATIONS IMPLEMENTING THE NUTRITION LABELING AND EDUCATION ACT OF 1990; OPPORTUNITY FOR COMMENTS; FINAL RULE
(21 CFR Parts 5, 20, 100, 101, 105 and 130) **[Docket No. 92N-0440]**

FOOD LABELING: ESTABLISHMENT OF DATE OF APPLICATION; FINAL RULE
(21 CFR Parts 5, 101, 105 and 130) **[Docket No. 90N-0134]**

FOOD LABELING: MANDATORY STATUS OF NUTRITION LABELING AND NUTRIENT CONTENT REVISION, FORMAT FOR NUTRITION LABEL; FINAL RULE *(21 CFR Parts 1 and 101)*
[Docket Nos. 90N-0135, 91N-0162, 78P-0091, 87P-0194/CP and 90P-0052]

Chapter XII

MANDATORY STATUS OF NUTRITION LABELING (continued)

MANDATORY STATUS OF NUTRITION LABELING (continued)

FINAL RULE

Chapter XII

FOOD LABELING; REFERENCE DAILY INTAKES AND DAILY REFERENCE VALUES; FINAL RULE
(21 CFR Parts 101 and 104) **[Docket No. 90N-0134]**

FINAL RULE

FOOD LABELING; SERVING SIZES; FINAL RULE *(21 CFR Part 101)* **[Docket No. 90N-0165]**

Chapter XII

FOOD LABELING; SERVING SIZES; FINAL RULE (continued)

FOOD LABELING; SERVING SIZES; FINAL RULE (continued)

FINAL RULE

FOOD LABELING: NUTRIENT CONTENT CLAIMS, GENERAL PRINCIPLES, PETITIONS, DEFINITION OF TERMS; DEFINITIONS OF NUTRIENT CONTENT CLAIMS FOR THE FAT, FATTY ACID, AND CHOLESTEROL CONTENT OF FOOD; FINAL RULE (*21 CFR Parts 5 and 101*) **[Docket Nos. 91N-0384 and 84N-0153]**

From the Office of Regulatory Affairs, FDA

Chapter XII

NUTRIENT CONTENT CLAIMS: Final Rule (continued)

Chapter XII

NUTRIENT CONTENT CLAIMS: Final Rule (continued)

FINAL RULE

FOOD LABELING: LABEL STATEMENTS ON FOODS FOR SPECIAL DIETARY USE; FINAL RULE *(21 CFR Part 105)* **[Docket No. 91N-384L]**

FINAL RULE

FOOD STANDARDS: REQUIREMENTS FOR FOODS NAMED BY USE OF A NUTRIENT CONTENT CLAIM AND A STANDARDIZED TERM; FINAL RULE *(21 CFR Part 130)*
[Docket No. 91N-0317 0317 et al.]

FINAL RULE

FOOD LABELING: USE OF NUTRIENT CONTENT CLAIMS FOR BUTTER; FINAL RULE
(21 CFR Part 101) **[Docket No. 91N-0344]**

Chapter XII

STATE PETIONS (continued)

FINAL RULE

CERTAIN MISBRANDING SECTIONS OF THE FEDERAL FOOD, DRUG, AND COSMETIC ACT THAT ARE, AND THAT ARE NOT, ADEQUATELY BEING IMPLEMENTED BY REGULATION; NOTICE OF FINAL LISTS; FINAL RULE
(21 CFR Part 1) **[Docket No. 91N-0134]**

Chapter XII

CERTAIN MISBRANDING SECTIONS: FINAL RULE (continued)

FOOD LABELING; GENERAL REQUIREMENTS FOR HEALTH CLAIMS FOR FOOD; FINAL RULE
(21 CFR Parts 20 and 101) **[Docket No. 85N-0061]**

GENERAL REQUIREMENTS FOR HEALTH CLAIMS FOR FOOD; FINAL RULE (continued)

FINAL RULE

FOOD LABELING: HEALTH CLAIMS AND LABEL STATEMENTS: DIETARY FIBER AND CANCER; FINAL RULE
(21 CFR Part 101) **[Docket No. 91N-0098]**

Chapter XII

HEALTH CLAIMS AND LABEL STATEMENTS: DIETARY etc.; FINAL RULE (continued)

FOOD LABELING: HEALTH CLAIMS AND LABEL STATEMENTS; DIETARY FIBER AND CARDIOVASCULAR DISEASE; FINAL RULE
(21 CFR Part 101) **[Docket No. 91N-0099]**

HEALTH CLAIMS AND LABEL STATEMENTS; DIETARY etc. (continued)

FOOD LABELING: HEALTH CLAIMS AND LABEL STATEMENTS FOLIC ACID AND NEURAL TUBE DEFECTS; FINAL RULE
(21 CFR Part 101) **[Docket No. 91N-0100]**

Chapter XII

HEALTH CLAIMS AND LABEL STATEMENTS FOLIC, etc.; FINAL RULE

FOOD LABELING: HEALTH CLAIMS AND LABEL STATEMENTS: ANTIOXIDANT VITAMINS AND CANCER; FINAL RULE
(21 CFR Part 101) **[Docket No. 91N-0101]**

Chapter XII

FINAL RULE

FOOD LABELING: HEALTH CLAIMS AND LABEL STATEMENTS; OMEGA-3 FATTY ACIDS AND CORONARY HEART DISEASE; FINAL RULE
(21 CFR Part 101) **[Docket No. 91N-0103]**

FINAL RULE

FOOD LABELING: HEALTH CLAIMS AND LABEL STATEMENTS; DIETARY SATURATED FAT AND CHOLESTEROL AND CORONARY HEART DISEASE; FINAL RULE
(21 CFR Part 101) **[Docket No. 91N-0096]**

FINAL RULE

FOOD LABELING: HEALTH CLAIMS AND LABEL STATEMENTS; DIETARY FAT AND CANCER; FINAL RULE *(21 CFR Part 101)* **[Docket No. 91N-0097]**

Chapter XII

FINAL RULE

FOOD LABELING: HEALTH CLAIMS AND LABEL STATEMENTS; SODIUM AND HYPERTENSION; FINAL RULE *(21 CFR Part 101)* **[Docket No. 91N-0095]**

FINAL RULE

FOOD LABELING: DECLARATION OF INGREDIENTS; FINAL RULE
(21 CFR Parts 101, 102, 130, 135, 136, 137, 139, 145, 146, 150, 152, 155, 156, 158, 160, 161, 163, 164, 166, 168 and 169) **[Docket No. 90N-0361]**

FINAL RULE

From the Office of Regulatory Affairs, FDA

Chapter XII

FINAL RULE (continued)

FINAL RULE (continued)

FOOD LABELING: DECLARATION OF INGREDIENTS FOR DAIRY PRODUCTS AND MAPLE SYRUP; FINAL RULE *(21 CFR Parts 131, 133, 135, and 168)* [Docket No. 90N-361D]

FINAL RULE (continued)

Chapter XII

FOOD LABELING: DECLARATION OF INGREDIENTS; COMMON OR USUAL NAME FOR NONSTANDARDIZED FOODS; DILUTED JUICE BEVERAGES; FINAL RULE
(21 CFR Parts 101 and 102) **[Docket No. 80N-0149]**

DECLARATION OF INGREDIENTS; COMMON OR USUAL NAME etc. (continued)

REGULATORY IMPACT ANALYSIS OF THE FINAL RULES TO AMEND THE FOOD LABELING REGULATIONS
[Docket No. 91N-0219]
REGULATORY IMPACT ANALYSIS STATEMENT *(21 CFR Parts 5, 20, 100, 101, 105, and 130)*

Prepared by: Don Aird, PAS Minneapolis, MN (2/11/93 corrected edition).
Edited/expanded: M-M. Richardson, PAS, St. Louis, MO
Re-formatted: P. Maroney-Benassi, Midwest Training Officer, Chicago, IL (5/93)

Minor editorial changes: J. Summers Associates, Inc. (3/97)

Chapter XIII
Index for Food Labeling Technical Amendments; August 18, 1993

INDEX FOR FOOD LABELING TECHNICAL AMENDMENTS; AUGUST 18, 1993

Food Labeling: Nutrient Content Claims, General Principles, Petitions, Definition of Terms; Definitions of Nutrient Content Claims for the Fat Fatty Acid, and Cholesterol Content of Foods; Food Standards: Requirements for Foods Named by Use of a Nutrient Content Claim and a Standardized Term; Technical Amendments; Final Rule
(21 CFR Parts 5 and 101), **[Docket Nos. 91N-0384, 84N-0153, and 91N-0317]**

Definitions of Nutrient Content Claims for the Fat Fatty Acid, etc.; Final Rule (continued)

FINAL RULE

PART 101---FOOD LABELING

FOOD LABELING: ESTABLISHMENT OF DATE OF APPLICATION; FINAL RULE
(21 CFR Parts 5, 101, 105, and 130) **[Docket No. 90N-134]**

FOOD LABELING; HEALTH CLAIMS: GENERAL REQUIREMENTS; FIBER-CONTAINING FRUITS, VEGETABLES, AND GRAIN PRODUCTS AND CANCER AND CORONARY HEALTH DISEASE; FRUITS AND VEGETABLES AND CANCER; AND FOLIC ACID AND NEURAL TUBE DEFECTS; TECHNICAL AMENDMENTS; FINAL RULE
(21 CFR Part 101) **[Docket Nos. 85N-0061, 91N-0098, 91N-0099, and 91N-0100]**

PART 101---FOOD LABELING

Chapter XIII

FOOD LABELING; SERVING SIZE; TECHNICAL AMENDMENTS; FINAL RULE
(21 CFR Part 101) **[Docket No. 90N-0165]**

FINAL RULE

PART 101---FOOD LABELING

FOOD LABELING: DECLARATION OF INGREDIENTS; COMMON OR USUAL NAME FOR NONSTANDARDIZED FOODS; DILUTED JUICE BEVERAGES; TECHNICAL AMENDMENTS; FINAL RULE *(21 CFR Parts 101 and 102)* **[Docket No. 80N-0140]**

FINAL RULE

PART 101---FOOD LABELING

Page	CFR	Topic
44063	101.30	Percentage juice declaration for foods purporting to be beverages that contain fruit or vegetable juice

PART 102---COMMON OR USUAL NAME FOR NONSTANDARDIZED FOODS

Page	CFR	Topic
44063	102.33	Beverages that contain fruit or vegetable juice

FOOD LABELING: MANDATORY STATUS OF NUTRITION LABELING AND NUTRIENT CONTENT REVISION, FORMAT FOR NUTRITION LABEL; TECHNICAL AMENDMENTS; FINAL RULE
(21 CFR Part 101) **[Docket Nos. 90N-0134 and 91N-0162]**

Chapter XIII

MANDATORY STATUS OF NUTRITION LABELING AND NUTRIENT CONTENT REVISION, FORMAT FOR NUTRITION LABEL; TECHNICAL AMENDMENTS; FINAL RULE (Continued)

FINAL RULE

PART 101---FOOD LABELING

Chapter XIV

Table of Contents: Chapter 5 – Foods, Colors, and Cosmetics for the Compliance Policy Guide

Adopted from Compliance Policy Guide, Chapter 5, Office of Regulatory
Affairs, FDA, Updated 4/20/01, and 11/29/05

	Adulteration by Insect and Rodent Filth; Mold; Mammalian Excreta	CPG 7109.26	11/29/05

Sub Chapter 527 - DAIRY

527.100	Butter - Adulteration Involving Insufficient Fat Content	CPG 7106.05	11/29/05
527.200	Cheese and Cheese Products - Adulteration with Filth	CPG 7106.07	08/96
527.225	Cheese - Misbranding Due to Moisture and Fat	CPG 7106.06	11/29/05
527.250	Cheese and Cheese Products - Misbranding Involving Net Contents	CPG 7106.09	11/29/05
527.300	Pathogens in Dairy Products	CPG 7106.08	11/29/05
527.350	Eggnog; Eggnog Flavored Milk - Common Usual Names	CPG 7106.02	11/29/05
527.400	Whole Milk, Low Fat Milk, Skim Milk - Aflatoxin M1	CPG 7106.10	11/29/05
527.450	Milk and Milk Products Containing Penicillin	CPG 7106.03	10/01/80
527.500	Malted Milk	CPG 7106.01	08/96
527.600	Use of DDVP (dichlorvos) Strips in Milkhouses and Milkrooms	CPG 7106.04	08/96

Sub Chapter 530 - DIETARY SUPPLEMENTS

530.400	Revoked: (FR Notice 9/23/97)	CPG 7121.02	
530.500	Wheat Germ Containing Non-Wheat Germ Tissue	CPG 7121.03	08/96

Sub Chapter 535 - EDIBLE OILS

535.100	Oleomargarine - Misbranding Due to Insufficient Fat	CPG 7113.01	08/96

Sub Chapter 537 - EGG INDUSTRY

537.100	Eggs and Egg Products - Frozen - Adulteration Involving Decomposition	CPG 7107.02	08/96

Sub Chapter 540 - FISH AND SEAFOOD

540.100	Capelin: Prohibited from Being Labeled as Smelt (Revoked 8/23/96)	CPG 7108.22	
540.150	Caviar, Use of Term – Labeling	CPG 7108.01	10/30/89
540.200	Chubs, Hot Process Smoked with Added Nitrite – Adulteration Involving Food Additives, Sodium Nitrite	CPG 7108.15	11/29/05
540.250	Clams, Mussels, Oysters, Fresh, Frozen or Canned – Paralytic Shellfish Poison	CPG 7108.20	11/29/05
540.275	Crabmeat - Fresh and Frozen - Adulteration with Filth Involving the Presence of the Organism Escherichia coli	CPG 7108.02	11/29/05
540.285	Crabmeat Products - Labeling; Crabmeat Products with Added Fish or Other Seafood Ingredients – Labeling	CPG 7108.03	10/01/80
540.300	Crabmeat - Product Name (Revoked 8/23/96)	CPG 7108.04	08/23/86
540.350	Common or Usual Names for Crustaceans (Revoked 8/23/96)	CPG 7108.23	08/23/80
540.375	Canned Salmon - Adulteration Involving Decomposition	CPG 7108.10	08/23/80
540.390	Canned Shrimp - Labeling, Size Designations and Corresponding Counts	CPG 7108.13	10/30/89
540.400	Shrimp - Fresh or Frozen, Raw, Headless, Peeled or Breaded – Adulteration Involving Decomposition (Revoked 12/24/96)	CPG 7108.11	12/24/96
540.410	Shrimp - Frozen, Raw, Breaded or Lightly Breaded – Misbranding Involving Non-Compliance with Standards	CPG 7108.12	11/29/05
540.420	Raw Breaded Shrimp - Microbiological Criteria or Evaluating Compliance with Current Good Manufacturing Practice Regulations	CPG 7108.25	11/29/05
540.450	Imitation Breaded Shrimp	CPG 7108.14	10/30/89
540.475	Snapper – Labeling	CPG 7108.21	10/01/80
540.500	Tuna, Sable, Salmon, Shad - Smoked Cured, Adulteration		

Adopted from Compliance Policy Guide, Chapter 5, Office of Regulatory Affairs, FDA, Updated 4/20/01, and 11/29/05

Chapter XIV

550.600	Olives - Adulteration Involving Pits; Rot; Insect Infestation	CPG 7110.19	11/29/05
550.605	Olives Stuffed with Minced Pimentos – Labeling	CPG 7110.20	11/29/05
550.625	Oranges - Artificial Coloring	CPG 7110.21	10/01/80
550.650	Peaches, Canned, Frozen - Adulteration Due to Insects and Mold	CPG 7110.22	11/29/05
550.655	Peaches, Canned - Misbranding Involving Food Standards	CPG 7110.23	11/29/05
550.680	Pineapple, Canned; Pineapple Juice - Adulteration with Mold	CPG 7110.24	03/95
550.685	Pineapple, Canned; Imported and Domestic - Misbranding Involving Food Standards	CPG 7110.25	11/29/05
550.690	Plums, Canned - Adulteration with Rot	CPG 7110.26	11/29/05
550.700	Dried Prunes, Dehydrated Low Moisture Prunes, and Pitted Prunes - Adulteration Involving Insects; Decomposition; Dirt; Pits; and Pit Fragments	CPG 7110.27	11/29/05
550.750	Raisins - Adulteration Involving Mold, Sand, Grit, & Insects	CPG 7110.28	11/29/05
550.800	Standardized Canned Fruit - Misbranding Involving Improper Declaration of Packing Medium	CPG 7110.29	11/29/05
550.850	Strawberries; Frozen, Whole, or Sliced - Adulteration with Sand, Mold	CPG 7110.30	11/29/05

Sub Chapter 555 - GENERAL

555.100	Alcohol; Use of Synthetic Alcohol in Foods	CPG 7120.10	02/01/89
555.200	Adulterated Food Mixed with Good Food	CPG 7120.14	10/01/80
555.250	Statement of Policy for Labeling and Preventing Cross-contact of Common Food Allergens	4/2001)	11/29/05
555.300	Foods, Except Dairy Products - Adulteration with Salmonella	CPG 7120.20	03/95
555.425	Foods, Adulteration Involving hard or Sharp Foreign Objects	(3/23/1999)	11/29/05
555.400	Foods - Adulteration with Aflatoxin	CPG 7120.26	11/29/05
555.425	Foods, Adulteration Involving Hard or Sharp Foreign Objects	(3/23/1999)	11/29/05
555.450	Foods - Adulteration Involving Infestation and 1080 Rodenticide	CPG 7120.02	10/01/80
555.500	All Food Sanitation (Including Bacteriological) Inspections – Classification of Establishment	CPG 7120.24	10/01/80
555.550	Foods, Standardized; Enriched or Fortified - Adulteration Involving Misbranding – Potency	CPG 7120.22	02/01/89
555.600	Interpretation of Insect Filth in Foods	CPG 7120.18	11/14/02
555.650	Reconditioning Foods by Diversion for Animal Feed	CPG 7120.21	03/95
555.700	Revocation of Tolerances for Cancelled Pesticides	CPG 7120.29	02/01/89
555.750	Seeds for Sprouting Prior to Food Use, i.e., Dried Mung Beans, Alfalfa Seeds, Etc.	CPG 7120.28	02/01/89
555.800	Polysorbates 20, 40, 60, 65, 80, 85 - Common or Usual Names	CPG 7120.09	02/01/89
555.850	Water Damaged Food Products in Screw-Top, Crimped-Cap and Similar Containers	CPG 7120.17	02/01/89
555.875	Water in Food Products (Ingredient or Adulterant)	CPG 7120.07	02/01/89

Sub Chapter 560 - IMPORTS

560.100	Importation of Unlabeled Foods - Exemption Under *21 CFR 101.100(d)**	CPG 7119.01	07/19/89
560.200	Country of Origin Labeling	CPG 7119.02	11/29/05
560.250	Imports - Importer can be Required to Reveal Identity of Ingredients	CPG 7119.03	07/19/89
560.300	Reconditioning of Imported, Insect Infested, Insect Damaged or Moldy Coffee Beans	CPG 7119.04	07/19/89
560.350	Coffee and Cocoa Bean Sweeps	CPG 7119.08	07/19/89
560.400	Imported Milk and Cream - Import Milk Act	CPG 7119.05	05/12/05

Adopted from Compliance Policy Guide, Chapter 5, Office of Regulatory Affairs, FDA, Updated 4/20/01, and 11/29/05

Adopted from Compliance Policy Guide, Chapter 5, Office of Regulatory Affairs, FDA, Updated 4/20/01, and 11/29/05

Chapter XIV

Chapter XV
Table of Contents for "Food Labeling Questions and Answers," August 1993

Adopted from FDA the publication "Food Labeling Questions and Answers" – for Guidance to Facilitate the Process of Developing or Revising Labels for Foods Other Than Dietary Supplements, August 1993

311

Adopted from FDA publication "Food Labeling Questions and Answers" – for Guidance to Facilitate the
Process of Developing or Revising Labels for Foods Other Than Dietary Supplements, August 1993

Chapter XVI

Table of Contents for "Food Labeling Questions and Answers," Volume II, August 1995

Adopted from the FDA publication "Food Labeling Questions and Answers", Volume II – A Guide for Restaurants and Other Retail Establishments, August 1995

313

Adopted from the FDA publication "Food Labeling Questions and Answers", Volume II – A Guide for Restaurants and Other Retail Establishments, August 1995

Index

Index

Index

Index

Bibliography

1. Compliance Policy Guide Manual, Chapter 5, Office of Regulatory Affairs, FDA, April 19, 2001, and Updated November, 29, 2005

2. FDA Guide to Nutrition Labeling and Education Act (NLEA) Requirements, August 1999 (Revised December 2005)

3. FDA Issued Letter to Industry on Foods Containing Botanical and Other Novel Ingredients, 2/5/01

4. Food Labeling: Trans Fatty Aids in nutrition Labeling, Nutrient Content Claims, and Health Claims, FDA, CFSAN, 8/2003

5. FDA Letter on Labeling Food Products Presented or Available on the Internet, Overview of Dietary Supplements, FDA, CFSAN, 1/03/01

6. Food, Drug, and Cosmetic Act, as amended, 1998

7. Food Labeling: Gluten-Free Labeling of Foods; Public Meeting; Request for Comments, Federal Register, Vol. 70, No. 137, July 19, 2005, pp. 41356-41358

8. Food Labeling: Questions & Answers, Foods Other than Dietary Supplements, FDA, August 1993

9. Food Labeling: Questions & Answers, Vol. II, A Guide for Restaurants and Other Retail Establishments, FDA, August 1995

10. Guidance for Industry and FDA: Requesting an Extension to Use Existing Label Stock After the Effective Date of January 1/2006, FDA, CFSAN, Issued December 2005, Revised 12/30/05

11. Guidance Letter for Industry "Notification of a Health Claim or Nutrient Content Claim Based on an .... Authoritative Statement of a Scientific Body," FDA, CFSAN, Office of Food Labeling (HFS-150), June 11, 1998

12. Guidance Letter for Industry Regarding Labeling of Bioengineered Products, U.S. FDA, CSFAN, January 2001

13. Guidance for Industry: Food Labeling: Safe Handling Statements, Labeling of Shell Eggs; Refrigeration of Shell Eggs Held for Retail Distribution (Small Entity Compliance Guide), FDA, CFSAN, ONPLD, July 2001

Bibliography

14. Guidance for Industry and FDA Staff: Whole Grain Label Statements; (For comments purposes only), FDA, CFSAN, February 15, 2006

15. Guidance on Labeling of Foods That Need Refrigeration by Consumers, Federal Register, Vol. 62, No. 36, February 24, 1997, pp. 8248-8252

16. "Health Claim Notification for Whole Grain Foods," July 1999 and "Health Claim Notification for Potassium Containing Foods," U.S. FDA, CFSAN, ONPLDS, October 2000

17. Letter about the Regulatory Status of the Good News Andy Pill, FDA, CFSAN, 5/30/00

18. Notifications of Health Claims Based on Authoritative Statements, Federal Register Vol. 64, 1/21/99; p. 3250

19. Other Evidence that the Intended Use of a Product is for the Diagnosis, Cure, Mitigation, Treatment, or Prevention of Disease, Title 21 Code of Federal Regulations, Section 201.128, Parts 200 to 299, 4/01/06

20. Title 21 Code of Federal Regulations, Parts 1 through 99, 4/01/06 Edition

21. Title 21 Code of Federal Regulations, Parts 100 through 169, 4/01/06 Edition

22. Title 21 Code of Federal Regulations, Parts 170 through 199, 4/01/06 Edition

23. Trans Fat Now Listed with Saturated Fat and Cholesterol on Nutrition Facts Label, FDA, CFSAN, ONPLDS, January 16, Updated 3/3/2004 and 1/1/06

24. The Dietary Supplement and Advertising Guide for Industry, FTC, BCP, 4/01/ and 7/16/03

25. "Understanding Food Allergy," International Food Information Council Foundation, Washington, DC, August 1998

26 U.S. FDA Talk Paper, "FDA Issued Letter to Industry on Foods Containing Botanical and Other Novel Ingredients," February 5, 2001